SHRINKING BIRDS

AND ELEVEN OTHER SHORT STORIES

FROM THE FRONTIERS OF BIOSCIENCE: 2016

JOHN R. SPEAKMAN

About the author and the articles

UK born, Professor John R. Speakman is a 1000 talents 'A' Professor at the Institute of Genetics and Developmental Biology, Chinese Academy of Sciences in Beijing, China, and a Professor in the Institute of Biological and Environmental Sciences at the University of Aberdeen, in the UK. He has lived in Beijing, China since 2011. In 2013 he also became a features writer for the Chinese monthly popular science magazine 'Newton'. The articles in this collection were written in English and were previously translated into Chinese and appeared each month in 'Newton'. This book is a compilation of the original English versions of the stories, modified for a western audience. The 'further reading' section at the back of the book lists the original sources on which the articles were based, and some suggested additional reading.

By the same author

Pandas – dead end or dead wrong? and eleven other short stories from the frontiers of bioscience 2014. Createspace publishing

How dogs make us fall in love with them and eleven other short stories from the frontiers of bioscience 2015. Createspace publishing

For Emily and Jack

Acknowledgements

I am grateful to Lina Zhang for translating these articles into Chinese and to Newton magazine for publishing the Chinese versions, while allowing me to retain copyright on the English versions of the articles. Thanks to Jan van de Kam for the royalty free use of the pictures of the birds on the covers and inside. The images of fat cells were kindly provided by Baoguo Li from the Molecular Energetics Group in Beijing

Cover and inside photo chapter 5:

    Jan van de Kam (Wildlife photographer)

        Deurneseweg 1
        5766 PH Griendtsveen
        Netherlands
        Jkam@worldonline.com

Photos chapter 8: Baoguo Li

Contents

January	Do animals keep fit?
February	Is democracy the best system of government, or the cradle of mediocrity?
March	Could killing this one type of cell make you live 25% longer?
April	Why do ant societies tolerate lazy individuals?
May	Shrinking birds
June	Women and children first, or last?
July	Why do we not feel hungry when we are ill?
August	Can we make our fat burn itself?
September	If you close your eyes how do you know where your legs are?
October	No more 'Bad air'

November	Smelling bee
December	Why do we treat the obese so badly?

Do animals keep fit?

January

We are lucky that our apartment in Beijing is right next to Olympic park. Located in the North of the city, and containing the Bird's Nest stadium and the Cube water park, this is the largest green area in the city. In some respects Olympic park is like any other large park in the middle of a large city, like Hyde Park in London or Central Park in New York. It has a large lake in it, and numerous trails where one can walk, but never get lost. In other respects however it is very different. We often go there on Sunday mornings to stroll around, and the first thing that contrasts large parks elsewhere in the world, that I have visited (except in China), is that on Sunday there is a large communal choir, singing away near the main park entrance. It is really nice to sit for a while and listen to the group of what must be a hundred or so people all singing in harmony. I never saw this on a regular basis anywhere except in China. But then something that makes the park the same as any other is that it is filled with people running or jogging. Indeed the park was set up before the Olympic games in 2008 with joggers obviously in mind. It has specific routes marked out that are 2, 5 and 10 km long winding their way through the trees and around the lake. My guess is that on a typical Sunday morning there are literally thousands of people jogging around these routes. They are not alone. Near to the park, and across the whole of Beijing there are hundreds of gyms that are full of people running, cycling and rowing to nowhere on exercise machines.

This behaviour of exercising, simply for the sake of exercising, seems to be a unique human activity. The park has lots of other animals in it apart from humans. But you never see groups of magpies flying in circuits around the trees. There are no synchronised groups of squirrels taking their morning 2km run through the tree tops. Even among our nearest relatives in the animal world nobody has seen groups of chimpanzees or gorillas running, or even walking, around tracks in the jungle just for the sake of it. Why? The classic explanation for this unique human activity is that for all other animals energy is in limited supply. Animals therefore need to spend much of their lives looking for food that will supply them with the energy they need to live. Very often if you watch an animal in the wild, what it is doing is feeding, or defending a piece of ground where it gets its food, from encroachment by others. Animals need to do this because their bodies have a relentless requirement for energy, and the only place they can get this is from food. If they are able to get more food than they need to cover their basic requirements then animals often allocate this extra energy to breeding. This view of the world inhabited by animals is pervasive in the field of ecology, and we can use it to understand why 'jogging' or equivalent exercise behaviours have seemingly never evolved in other animals.

Consider for a moment an animal that had a mutation that caused it to engage in a regular but pointless exercise like jogging for an hour every day. Let's call it the JOG mutation. Such exercise requires energy. More energy than just sitting around doing nothing. Compared to an animal without the JOG mutation, animals with the mutation would consequently need to gather more food, and hence feed for longer each day. But the time they have available for feeding would be correspondingly 1 hour shorter, because of the time they had spent running. If energy was in short supply, as is generally presumed, then individuals carrying the JOG mutation would be at much more risk of failing to make a balance in their energy budgets. And, if you repeatedly fail to balance your energy budget by not matching energy intake with energy expenditure, then there is an inevitable result – death. So the JOG mutation would enhance the risk of dying. That isn't fatal for the spread of a mutation if the risk of mortality is more than offset by an increase in the rate at which carriers of the mutation produce offspring. However, for the JOG mutation that seems unlikely. If we imagine enough food is available to survive, whether individuals have the JOG mutation or not, non carriers would be at an advantage because they would have more energy available to produce babies, because they didn't waste any of it on jogging. The JOG mutation would be very unlikely to spread in the population because it would lower both survival and fecundity. It would always be outcompeted by the non-JOG alternative.

Humans seem to have overcome this problem because for many people in the modern world energy is no longer limited. Once you become released from the restricted energy supply then it becomes feasible to engage in lots of things that are not feasible for other animals. This is why humans can spend their time and energy doing all sorts of things like art, science, religion, war and politics that other animals cannot do because they need to devote all their time to feeding and breeding. Jogging, and other forms of pointless exercise, are just another thing to add to the list of what we can do because we are not energy limited. In fact we know that propensity to exercise has a genetic component to it. Is it possible that the JOG mutation spread in human populations because there is no energy limitation to stop it? There are lots of things that are consistent with the energy limitation interpretation. First, engaging in this type of exercise is strongly related to how rich you are, and hence how much money you can spend on food. That is, how released you are from the constrained energy supply. Take yourself outside Beijing to a rural village, and my guess is that on Sunday morning the joggers will be few and far between. Plus you would probably be hard pressed to find a gym full of exercise machines. Second, such exercising is a relatively modern phenomenon. As we have grown rich and released ourselves from the yoke of limited energy, spending energy on exercise has become feasible. If you read an old text like the bible, then at no point do you ever come across a sentence like 'One day Moses was out running…' or 'When Jesus went out for his weekly jog he noticed… etc'. I have never read Confucius but my bet is that if you do you will find the same thing. I couldn't find online any of his famous quotes that concern engaging in pointless exercise (good or bad). Two to three thousand years ago people just didn't do this, presumably because they did not have the spare time or (energy) resources to allocate to it. It appears that such exercising has appeared only in the last 400 years or so. Jogging as a word appears in English literature around the mid 1650's. So everything seems to make evolutionary sense. I can look on at joggers in the park slogging around the 10km track, smug in the interpretation that they are defective mutants with the JOG gene who would normally have been eliminated by natural selection, but can survive now in our benign non-energy limited modern human society.

But does that evolutionary story really make sense? Five years ago myself and a colleague Ela Krol, published a paper in the Journal of Animal Ecology, in which we aimed to challenge the long held idea that energy supply for most animals is limited. There were several stands to the argument but here are a couple of things that we pointed out. The first is that if you experimentally remove the limited food supply then animals do not generally respond in a manner suggesting they were previously limited. There have been many experiments in which the food supply to wild animals has been supplemented. When this is done what happens is the animals do not suddenly eat more food and spend the extra energy engaging in greater levels of reproduction. The most common response is that the animals eat the same amount of food, but they reduce the time they spend foraging to attain it. This suggests something limits their ability to expend energy rather than acquire it. Another thing to think about is this. If you compare the energy needs of say a lactating mouse and a non-lactating mouse, then we know that during lactation the energy demands are about 5x greater, and hence the food intake is correspondingly also about 5x higher in the lactating animal. During the breeding season the population consists of animals at various stages of reproduction. The lactating animals are normally able to find enough food to supply their enormous energy demands. For a non-lactating animal in the same environment therefore there must be considerably more food available than it requires. It is impossible to imagine that all individual animals in a population are limited by energy supply, when their energy requirements are so variable.

Although we did not make this argument in our paper, it has some important implications - if most animals are not energy limited most of the time, then the reason they do not exercise, like humans do, cannot be because of limited energy supply. In the last month (January 2016) Lewis Halsey, a biologist from the UK, has published a paper, also in the Journal of Animal Ecology which has blown this whole field wide open. Halsey has made the revolutionary suggestion that in the wild many animals probably do exercise like humans do! How is that possible? Why don't we see them? What about the evolutionary argument that we made above concerning the JOG mutation, and how it would always be outcompeted by individuals not carrying such a mutation. Halsey asks us to consider a different scenario. What is the consequence of all that jogging? One thing that is very clear is that people who do engage in such exercise are 'fitter' than those who don't. Here we are not talking about evolutionary fitness, but physical fitness. This has consequences for physical performance. So if you were to pick at random anyone running around Olympic park on a Sunday morning, and then put them into a 200m running race against me (who hardly ever exercises), they would definitely always win. It is only a small extension to this scenario to consider the evolutionary consequences. What if we were being chased by a tiger? This reminds me of the old story about two visitors to Australia talking about being chased by a crocodile. One of them says to the other "I don't think you could outrun a crocodile". The other replies, "I don't need to. I just need to outrun you!"

The point is that 'pointless' exercise is not always (or even normally) evolutionarily disadvantageous. Animals that exercise may expend more energy, but if energy isn't limited (see the arguments above from our paper) then this is unlikely to routinely lead them into a situation where they don't make an energy balance, and die. The downside may be minimal. Yet it could have some dramatic advantages if they ever encounter a predator, because then carrying the JOG mutation may mean the difference between surviving and breeding, or being tiger food. In that situation one might easily imagine that the JOG mutation would rapidly spread in the population. For an animal that encounters predation risk, physical fitness may lead to evolutionary fitness. It is a paradigm shifting perspective.

So why do we not see animals running around parks on Sunday mornings? One possibility is that often animals may do enough exercise simply by engaging in their routine foraging behaviours. They may only need to do extra exercise when the levels they routinely engage in are insufficient. This could be a potential reason why jogging only appeared in humans in the last few hundred years, and then only in richer segments of society. Prior to that, and in rural communities, with many people engaged in physically demanding occupations, extra exercise may not have been necessary from a fitness perspective. However, the question is, how much is enough? There may be some evolutionary advantage in fitness terms by doing a bit more than the routine. The key problem is would we actually be able to identify such a behaviour if we saw it. For example, if I sit in the park and watch a magpie fly from one tree to another, how would I know if this flight had any other purpose than the animal doing the flying to keep itself fit? I don't interpret it as jogging like exercise because I am primed by the assumption that animals don't do that sort of thing. However, because animals don't run laps does not mean they don't exercise to keep fit. When we observe animals in the wild we maybe need to recalibrate what we are seeing, through the different lens provided by Halsey's paper. Animals may not run on exercise equipment in the wild, simply because the equipment isn't there! What if we were to put exercise equipment out for them? Would they use it? In fact last year there was an interesting study where scientists did just that. They put rodent running wheels out into the field and watched what happened. Amazingly they found that wild rodents actually used them! Is this evidence that animals do routinely exercise just to keep fit?

Now when I go to the park I see things in a whole new light. Birds apparently aimlessly flying from tree to tree may be ensuring their physical fitness for the inevitable encounter with a hawk. And what about the joggers? As I noted above, we know that there is at least some genetic component to the propensity to exercise. Do the people running in the park on Sunday simply still have the genetic machinery that was selected in our past when we were susceptible to predation? Are they preparing for the encounter with a tiger that will never come. Have the rest of us, that do not exercise, lost the genetic machinery that impels us to do it, because since we eliminated the risk of predation it was still possible to survive without the JOG mutation? Are we non exercisers in fact the defective mutants that would normally have been eliminated by natural selection, but are able to survive now because modern human society lacks predators?

Is democracy the best system of government: or the cradle of mediocrity?

February

When the cold war ended in the early 1990s, the guiding principal of foreign policy in the USA switched from containment of the 'Warsaw pact' countries, dominated by the old Soviet Union, to the international spread of democracy. As you will know, democracy is the system of government in which the ruling body of the nation is elected by the people. It is derived from the Greek word *demos* meaning 'of the people', and as a form of government it has its origins in ancient Greece and the Greek city states such as Athens. The idea behind a democracy is that the people vote for who they want to lead them. This has the effect of ensuring that the decisions made by the people in power, reflect the wishes of the population in general, because in theory, people would not vote into power someone who believes and promotes something the people do not want. Moreover, if someone turns out to make decisions the people do not agree with, then by the process of democratic elections those leaders can in theory be replaced. Although the number of people who have the ability to participate in such electoral processes has varied over time, in most modern 'democracies' the franchise to vote is normally extended to all citizens of a nation, irrespective of their gender, race or beliefs – called 'universal suffrage'. It is widely believed, by those people that promote democracy, that it is the best possible form of government, that it ensures the fair election of the very best leaders, who have the best interests of the people, as opposed to their own best interests, at heart. Consequently, it is believed that it should be promoted as a model system that should be adopted by all other countries. Winston Churchill famously said *"Democracy is the worst form of government. Except for all the others."* Democracy is strongly tied up with the idea of political freedom, because it is argued that without the freedom to express ideas, then it is impossible to have a valid democratic debate on which to base a democratic voting process. It is an axiomatic belief in countries that have democracy that it is something that is wanted by people living in countries that do not have it. The foreign policy mission of the USA to spread democracy through the world has been pursued with an almost evangelical zeal.

Most Americans, and members of other democratic societies, assume that democracy is a good thing, and that the spread of democracy is also a good thing. The virtues of democracy are taken for granted, and it is widely suggested that the spread of democracy will improve the lives of citizens in countries where it is adopted, promote economic growth, reduce violence and contribute to the spread of international peace. In fact there is some objective evidence that democratic governments are less likely to kill their own people. Work by R.J. Rummel (in his book, "Power Kills: democracy as a method of non-violence", 1997) showed that on average members of democratic states, between 1900 and 1987, had only a 0.14% chance of being killed by internal violence, which was compared to 0.59% for authoritarian states, and 1.48% for totalitarian states. It is argued that this is because a government killing its own people would not survive the democratic process and remain in power. However, while democratic states may be relatively benevolent to their own populations, presumably because of the risk of being deselected from office, this largess is seldom extended to members of other states: presumably because members of other states cannot vote in their elections. Indeed, it is democracies that have the worst record in this respect. One need only think of the civilian bombing campaigns of the USA against Japan in world war 2 (including the only use of nuclear weapons against civilian populations to date), and the carpet bombing of Cambodia by the USA in the 1960s, during which over 560,000 tons of bombs were dropped, as prominent examples of how democratic governments do not respect the lives of civilian citizens of other countries that cannot vote them out of power. The net effect of 'democracy' on total mortality rates of non-military personnel is therefore unclear.

Another common argument made is that democracy leads to greater economic prosperity. It is true that currently the largest economy in the world is a democratic country. But the correlation between democracy and economic prosperity is weak. Indeed, estimates vary, but few analysts doubt that China will become a larger economy than the USA sometime between 2025 and 2050. Indeed by some measures used by the International monetary fund (e.g. GDP at purchasing power parity which takes into account relative prices as well as GDP itself) the economies of the USA and China are already about the same size. What then for the argument that democracy is a pre-requisite for economic prosperity? Indeed it has been often observed that totalitarian dictators can sometimes enact good economic policies that lead to impressive growth (see for example the book *Dictatorship, Democracy and Growth* by Mancur Olson).

So why am I telling you all this political stuff in an article that is supposed to be about bioscience and evolutionary psychology? The reason is that people (in the West) have rarely questioned the idea that democracy is a good thing, but now evolutionary psychologists are starting to study some of the basic ideas that underpin political systems, and how they evolve, and the results are quite interesting in the context of which political systems generate the best leaders.

A fundamental assumption that underlies the democratic process is the notion that the voting public have the competence to recognise the best policy (or the most able leader) from among the available candidates. This is why for example patients that are held in institutions for treatment of mental illness are not permitted to vote in most democracies, because it is presumed they do not have the mental capacity or stability to make such a rational decision. An expanding body of work, however, suggests that the competency to recognise the best policy or candidate may be far more widely absent in the general population. In fact, the idea that on average most populations are pretty stupid, and hence they should not be entrusted with the important job of deciding who should be the leader, has been hinted at by many prominent observers. Winston Churchill, in his second famous quote about democracy said *"The best argument against democracy is a five-minute conversation with the average voter."* and the American journalist Henry Louis Menken said of the US system *"As democracy is perfected, the office of president represents, more and more closely, the inner soul of the people. On some great day...the White House will be adorned by a downright moron."* Isaac Asimov, the science fiction writer, has bemoaned the fact that democracy has fuelled a sort of anti-intellectualism, around the notion of equality of thought, when he stated that democracy has come to mean *'my ignorance is just as important as your *knowledge'*.

The problem is not just that large sectors of the population are incompetent, but that people are also self delusional when it comes to rating their own abilities and intellectual skills. This is called the 'Dunning-Kruger' effect after two psychologists at Cornell University in the USA who first described it in 1999. The effect was discovered in a series of experiments that were inspired by a 'genius' bank robber in the USA who had covered his face with lemon juice before committing the robberies. The rationale for his behaviour was that lemon juice is used in invisible ink, and so if he covered his face in it, it would prevent his face being recorded by CCTV cameras in the banks he robbed! This led to the idea that people could be so incompetent that they couldn't even evaluate how incompetent they were! They tested this idea on Cornell undergraduates by giving them a range of tests which included logical reasoning, appreciation of humour and grammar tests. Once the tests were completed the students were shown the range of scores that the whole class had achieved, and were asked what they thought their own performance was. Students who were competent at the skills generally ranked themselves correctly. However, people who were incompetent mistakenly rated themselves as being really good. For example, even individuals that were in the bottom 12% of scores on a given test, consistently ranked themselves as in the top half performance wise. In other words, everyone thought they were 'above average', whether they were or not. However, not only did incompetent people fail to recognise their own incompetence, they were also seemingly unable to recognise genuine skill in others. In another study Kruger and Dunning got students to grade a quiz that tested grammar skills. They found that students who had done poorly on the test themselves gave poor marks to students who had actually performed well. That is they not only got the answers wrong, but they couldn't identify the correct answers when presented with them. This is really not so surprising if they thought they had the right answers to start with. For the resulting paper entitled 'Unskilled and unaware of it' that was published in the Journal of Personality and Social Psychology in 1999, Kruger and Dunning were awarded one of the 2000 Ig-Nobel prizes. [For those of you that are unaware the Ig-Nobel prizes are a parody of the Nobel prizes, and are awarded annually for the ten most trivial achievements in science published during the previous year.]

Being awarded an Ig-Nobel prize seems a bit harsh, because it turns out the Dunning-Kruger effect has some important ramifications when it comes to voting systems, and the ability of the democratic process to select the best leader. Put simply, if incompetent people are unable to judge the competence of other people, or the quality of the ideas that other people have, then allowing them to vote must surely undermine the ability of a democratic process to generate the best leaders, or the best solutions to important problems. For example, if people lack expertise on climate change, then it is difficult for them to identify people who actually are experts, or to evaluate the problem and the merits of suggested solutions. Because such people lack the mental tools to evaluate complex information, providing additional information about an issue, or facts about the abilities of different candidates, will have no effect on their decision making.

The impact of the Dunning-Kruger effect on the democratic process was actually evaluated more formally in 2010 by a German medical doctor called Mato Nagel, who is a specialist in kidney disease. Nagel simulated the impact of the Dunning-Kruger effect on election results by assuming that the population consists of people that vary in what he called a competency quotient (CQ) that follows a normal distribution with a mean of 100 (a bit like the intelligence quotient or IQ). The task he set is that the population must elect candidates to be leaders, from a pool of candidates that are drawn from the general population, and hence also vary in the same way in their CQ. If the democratic process works well it should select the person with the highest competence from the pool of candidates. To simulate the 'Dunning-Kruger' effect he assumed that voters could not distinguish between candidates that had a higher CQ than their own CQ. For example, if someone had a CQ of 50 they would be unable to distinguish between candidates that had a CQ above 50 and hence might select any of these candidates at random, thinking they were the best. If a person had a CQ of 150, however, they would be able to identify the CQ of most of the candidates, and hence be much better at selecting the best one. When Nagel simulated democratic elections using these parameters the result was that normally elections did NOT select the most competent candidates, but rather selected individuals who were barely above the population average. Democracy it seems is the cradle of mediocrity. The interesting thing is that because these elected people would be unable to distinguish how good they actually were, relative to much better candidates, they would consider themselves the best, and their election would be seen as vindication of this feeling! In fact this phenomenon was captured nicely in the quotation by the author Alain de Botton when he said *"Politics is so difficult, it's generally only people who aren't quite up to the task who feel convinced they are."* The main strength of democracy then is not that it produces the best leaders, but that it avoids the worst. Even if it is assumed that in authoritarian and totalitarian systems leaders emerge at random, relative to their competence, then about half of the time the leader of such a regime would be more competent than a leader generated by the democratic process. For some reason this result isn't being shouted from the hilltops by the Western democracies!

For someone from the West who has been exposed their whole life to the idea that democracy is a good thing, that produces the best solution to this sort of problem, this came as a bit of a shock to me. Consequently, when I read this paper I wondered how the democratic system might be improved to ensure that it selects the best leader. In fact one way to improve it is obvious. You need to consider only the voting decisions of the most competent people. But that leads to a problem because the notion of 'universal suffrage' – the right of everyone to vote, is an integral part of modern democracy. A less harsh solution would be to let everyone vote, but then when you count the votes place more weight on the votes by the most competent individuals. This creates another problem, however, and that is how you decide who is most competent. But actually given modern computer technology there is an easy way to do that. You could do it all electronically, by having a bank of say 10,000 questions that test how competent people are. When you go to vote, before you make your decision between the candidates, you would need to answer 10 of these questions that the computer would select at random from the 10,000 available. Your vote would then be multiplied by the number of questions you got right. This way the voting decisions of highly competent people would be weighted the highest. The key to success of such a system would be that the actual scores of people on the test should never be revealed. That way because of the 'Dunning-Kruger' effect everyone would think they had performed above average on the competency test, and that their opinion was being considered more important than that of most other people. Everyone goes home happy, and the system would 'democratically' select the most competent leader.

Follow-up

In the wake of the Brexit vote in June 2016, many voters on the losing 'remain' side claimed that the people who voted to leave the EU were generally ignorant and uninformed of the consequences of their votes. If that was true then a system like that advocated above might have resulted in a different outcome. However, I am not convinced that voters in the 'remain' camp were actually any more informed than the leavers. So in reality it may have made little difference.

Could killing this one type of cell make you live 25% longer?

March

Three years ago this coming July my mother died. She outlived my father by 4 years. When your first parent dies it is traumatic. But when your second parent dies it is much worse because then you have a difficult task to perform, and that is to go through all the things that they accumulated in their lives and decide what is valuable, what is sentimental and what is to be thrown away as junk. For my family it took us some time before we could build ourselves up to it. Then came the day when I went with one of my sisters to the house where my parents had lived since the second world war, and where we had both grown up through our entire childhoods to do 'the big clear out'. It was a small modest house. But it was packed full of things that they had accumulated over the sixty years they had lived there. It became very clear that my parents liked to keep hold of things that they should probably have thrown out many years previously. In what had formerly been my bedroom, and long since converted into an enormous storage cupboard, we found a hoard of vacuum cleaners dating back to the 1950s. All broken, and waiting for the day that they might be repaired. In the garden shed we found a tin helmet from the second world war. Hanging up on a hook, waiting for another potential invasion. I discovered all the notebooks that I had ever used throughout my whole time in secondary school. Dutifully waiting for the day I might return and need to know what the formula for the volume of a sphere is, or remind myself of the causes of the French revolution. Then there were pile upon pile of motoring magazines from the 1930s with instructions on how to repair cars that had gone out of manufacture before I was born. At first it was upsetting, especially for my sister, as we heaved things into bags to take to the tip and throw away. But it was clear we couldn't keep all this stuff, and accumulate it into our own domestic collections, waiting for our own children to sort through when the inevitable time comes. So then it became a fascinating journey into their lives that we could never have done when they were alive.

As time had gone on, particularly after my sisters and I had left home, my parents had held onto more and more things so that their house gradually became fuller and less functional as a home. As noted above, my bedroom became an enormous cupboard, and my sisters' room became steadily less of a bedroom and more of a storeroom that incidentally had a bed in it, somewhere behind the piles of old magazines. However, their hoarding behaviour was relatively minor compared to the behaviour of some people. Indeed extreme hoarding was classified in 2013 as a psychological mental disorder akin to obsessive compulsive disorders. In fact there was a TV show in the USA that highlighted some of the lives of extreme hoarders that first aired in 2009. It is currently in its 8th season, which started in January 2016. The longevity of the show highlights the fascination that people have with such a mental disorder. And if you see the show it is truly amazing what the homes of people who have this hoarding disorder are like. They make my parents home like a paragon of simplicity and minimalism. Rooms are filled to the brim with newspapers, and quite often bags of garbage that the person couldn't bring themselves to let go of. Perhaps one of the most extreme examples, however, was the home of a man living in Newton, Connecticut in the USA that was not only filled with magazine and newspapers, and bin bags of rubbish, but also contained almost 300 five litre plastic tubs full of urine (about 2 years worth). I say 'one' of the most extreme because there are worse stories out there! The disruption of everyday life in my parents home was minor.

In the home of a compulsive hoarder normal life becomes pretty much impossible.

In a strange way our bodies are a bit like our homes in that over time our bodies also accumulate junk. The junk in this case are cells that have become senescent. Over time the cells in our bodies are continuously exposed to stress and damage from a variety of sources. There are two extremes in the responses that cells have to such stress. One is that they recover completely. The other is that they die and are removed. However, there is a third less common option and that is cells which normally divide (called proliferating cells) can stop proliferating and enter a state of permanent arrest. In this state they are called senescent. In fact senescence was discovered half a century ago by a researcher called Leonard Hayflick who was working at the University of Texas in Galveston. Hayflick observed that when cells were grown in culture dishes they initially divided rapidly, but then growth slowed and eventually the cells stopped dividing completely. This occurred even though the cells were still alive and otherwise seemed to be functioning normally. Since there seemed to be abundant nutrients still available to support growth, Hayflick postulated that cells are only able to divide a fixed number of times, after which they become senescent. At the time this was a revolutionary suggestion. So revolutionary that it was almost impossible to publish in a scientific journal. In 1999 I went to a meeting in the UK where Hayflick was being presented with the Lord Cohen medal from the British Society for Research on Ageing. In his acceptance speech Hayflick showed pictures of all the rejection letters he had received when he tried to publish the idea that the number of cell divisions is limited. Following one memorable rejection the editor of the journal actually wrote to Hayflick's then head of department recommending that he be sacked for his sloppy work! Eventually in 1965 the work was published in the journal *Experimental Cell Research,* which is a respectable journal but by no means among the top flight. Hayflick had the last laugh. His paper has been cited over 3600 times by other scientific papers, and it is among the top 100 cited papers from over 2 million research papers published between 1961 and 1978. The number of cell divisions that a cell can make before becoming senescent has since become known as the 'Hayflick limit'.

Cells in our bodies may senesce because they reach the 'Hayflick limit' but a more common reason is because they get damaged. It is thought that a reason cells may become senescent is to prevent themselves developing into cancers. Senescent cells that have stopped dividing are a bit like the junk in our homes. They hang around clogging up the place. As we age the numbers of senescent cells in our bodies gets larger in the same way that our homes generally acquire more and more memorabilia of our passing lives. Estimates are that up to 15% of cells in aged tissues may be senescent. However, there is an important difference. It appears that senescent cells don't just sit there like our junk does. They actually become resistant to signals that would cause them to die, and they show a profound change in physiology, which includes an upregulation of genes that encode many secreted proteins. This secretory behaviour was discovered by several groups in the early 2000's, chief among which was the group led by Judith Campesi from the Buck Institute in California. The cocktail of secreted proteins was called SASP (the Senescence Associated Secretory Phenotype). These SASP proteins can have a large impact on the function of nearby non-senescent cells. In particular this cocktail of secreted proteins may disrupt normal cells and stimulate the growth of premalignant cancer cells. This is somewhat ironic given one of the main ideas for why cells might go into a senescent state in the first place is to prevent themselves becoming cancerous. So senescent cells may themselves be unable to replicate and become cancers, but they may fuel the development of cancer among their near neighbours. Senescent cells appear far from just benign passengers clogging up our tissues as we get older. They actually may contribute to, or be the main cause of, ageing itself.

An obvious question then is what would happen if we were to simply clean out all the senescent cells in our bodies? Would we live longer? A team led by cancer specialist, Professor Jan van Deursen, from the Mayo Clinic College of Medicine in the USA, have just completed such an experiment in mice, and the results were published last month (February) in *Nature*. The first issue they faced is how to kill just senescent cells. The way the authors did this was to genetically engineer mice by inserting a gene that was activated when the mice were injected with a drug called AP20187. Activating this gene caused the senescent cells to self destruct. The experiment involved keeping mice until they were middle aged (1 year old in mice) and then injecting some of them with the drug twice a week. Although this wasn't 100% effective it did reduce the numbers of senescent cells by about 60%. The impacts however were significant. In mice that received the drug, causing their senescent cells to die, there were significant improvements in the functioning of their hearts and kidneys. Plus behaviourally the treated mice behaved like much younger individuals – being more active and showing more exploring behaviours in novel environments.

Plus they lived longer. The tests were performed in two strains of mice and in both sexes. The lifespan increase in the first strain averaged across both sexes was from 624 days in mice where the senescent cells were not cleared to 793 days in those where they were. An increase of 27%. In the second strain the values were 699 days in the un-cleared mice and 866 in the ones with the senescent cells cleared. An increase of 24%. Even van Deursen in a press interview afterwards said he was surprised at the life extension, given they had only removed 60% of the senescent cells.

To put that into a human context, the average life expectancy in the USA is about 80. If the same 24-27% effect was replicated in Americans then life expectancy would increase to about 100 years. Figures for most modern states are similar to those for the USA. Of course, having extra years isn't everything. I once gave a talk about scientific strategies for life extension to the general public at a Science festival in the UK. At the end someone in the audience made an important and astute observation. "The problem" he said "with life extension, is that the extra years are always added onto the end. But these are extra years that nobody really wants. What we all want are extra years 'in the middle'.". The important thing about the life extension in the mice with reduced senescent cells is that they also showed some improvements in their health and behaviour. However, some things that change with age were not impacted. Their memory, muscle strength, co-ordination and balance were no better than the untreated animals of the same age. So the extra life was not exactly added 'in the middle', but it was also not entirely 'at the end' either.

How realistic is the goal of clearing out the junk senescent cells in humans? After all the mice had to be specifically genetically engineered so that a drug could then target these cells specifically. Apparently this may not be too far off. Van Deursen, the leader of the mouse study that was just published in *Nature*, is co-founder of a company called 'Unity Biotechnology' (www.unitybiotechnology.com) that was launched on the same day that the *Nature* paper was published, with the goal to target ageing via depletion of senescent cells. Interestingly, Judith Campesi who, as noted above, was one of the first scientists to discover the SASP phenomenon, is also a scientific cofounder. Apparently the company already owns the intellectual property for a portfolio of drugs that can kill senescent cells in normal mice that have not been genetically engineered. If these prove to have a favourable effect on life and healthspan in mice, then preliminary clinical trials in humans cannot be too far away. Clearing out the junk may be the best strategy yet for adding life 'in the middle'.

Why do ant societies tolerate lazy individuals?

April

At some point most managers have looked at tables of the performance of workers in their own area of responsibility and noticed that performance seems to follow a distribution where there are a few people doing very well, followed by a large numbers that have an average level of performance and bringing up the rear a tail end of individuals who perform well below average. Of course, by definition, some individuals have to be below average, because that is how the average is calculated. Nevertheless it is clear that if you could get rid of these underperforming individuals then the average performance of the team as a whole would increase. Getting rid of the tail has become firmly entrenched as an idea for improving overall unit performance in management circles. We generally imagine that the performance tail is something unique to human activities like organised work. This is presumably because in human society we have a benevolent attitude when it comes to individual variation in performance. Managers may dream of chopping off the tail, but in reality this is seldom done. We tolerate underperformance and sustain harmony by paying people different amounts according to how hard they actually work. So lazy Dave who always slopes into the office last with some excuse that the bus was delayed by the traffic does not create too much disharmony, because when it comes to bonus time his rewards are substantially lower than those of super productive Jane who is always first into the office, and she is always the last to leave.

This sort of system, one might imagine, could never evolve by a process of natural selection in, for example, insect societies. Imagine for example an insect society like a colony of bees or ants, where this happened. That is where the colony consisted of some crazy diligent individuals who go out each day searching for food and generally contributing positively to the functioning of the society. But in addition to these individuals there is also a group of individuals that just hang around the colony doing not very much, only occasionally going out to forage, but most of the time just hanging about eating their way through the colonies shared resources. Then imagine another colony that consisted only of the type of individuals that worked amazingly hard and always contributed to the colony whenever there was a task required, in addition to utilising the shared colony resources like food. What would happen if these two colonies suddenly hit up against a challenge like a drop in food supply? One could easily imagine that the colony comprised only of dedicated workers would send out everyone looking for food and hence their chance of finding food and surviving would be much greater than the colony where only half the individuals went out and the others sat at home munching through what little food they had. Colonies of exclusively hard working individuals one would imagine should always have an edge over those comprising a mix of 'workers and shirkers' if natural selection and survival of the fittest governs the way these systems evolve. Indeed mathematical optimality models predict exactly this outcome. Insect societies should consist of overwhelmingly dedicated workers because short term increases in work rate of any one individual contributes positively to the evolutionary fitness of the whole colony. Hence colonies consisting only of hard workers are predicted to have greater survival, and produce more offspring that will go off and form their own colonies, taking with them the 'hard work' genes.

It might come as a surprise then to find that actually many ant colonies appear to more often consist of a mix of individuals that vary enormously in their work rates and diligence. In fact at any one time almost 50% of individual ants may be inactive, and some individuals show an extreme reluctance to take part in doing any of the work. How is it possible that natural selection has favoured the evolution of a system in insect societies that seems remarkably similar to the world of work for humans? Why do colonies of super hard workers not always evolve? Why do the hard workers not throw the lazy ones out? Researchers on ants refer to these differences between individuals in a given colony as variation in their 'task response threshold'. That is individuals vary in how wiling they are to respond to the need for a task to be done. The hard workers have a low task response threshold – so they work all the time because there is always something to be done that is above their threshold. You may know someone who has a low task threshold. They are the sort of person who is perpetually tidying up and cleaning even when they have just done it ten minutes previously. On the other hand individuals with a high task threshold fail to respond to almost all tasks.

One interpretation is that there is perhaps some division of labour going on. The individuals with high thresholds to perform are not doing the most obvious tasks but they contribute in some other less conspicuous way to the overall productivity and colony functioning. Indeed they may be actually incapable of doing these more obvious tasks which is why they are not observed doing them. But this seems not to be the case. Experiments have been done where a colony was observed and the lazy individuals are quite happy doing nothing at all except watch the hard workers get on with it. They don't seem to be doing any covert essential functions. Plus, if the hard workers with low thresholds are then removed the lazy ones with high thresholds start to do the work. So this isn't a matter of them being unable to work. They just have a high threshold. Anyone who has raised children to adulthood will have experienced this first hand, when they hit their mid-teenage years. They are not incapable of working – they just have a high task response threshold.

An interesting question then is how insect societies consisting individuals with low and high thresholds have been able to evolve, when a constant low threshold across all individuals would seem to always win in terms of productivity. A team mostly from Hokkaido University in Japan, but including several other universities in Japan, led by Eisuke Hasegawa and Jin Yoshimura have addressed this perplexing issue in a paper published recently (February 2016) in the journal *Scientific reports*. The authors posed an interesting scenario that might explain why mixed ant colonies evolve. They noted that not even the really hard working individuals can work all the time. They fatigue when performing tasks and need to rest. They also noted that there are a set of tasks in all the colonies that need to be done perpetually to ensure the long term persistence of the colony. Things like caring for eggs and larvae are absolutely essential, and studies have shown that even if these get interrupted for only short periods then viability of the eggs and larvae is reduced. Imagine now there was a catastrophic event that hits the colony. What would happen? In a colony consisting of all hard workers they would all go to work solving the catastrophe which would be above their work threshold but then they would potentially all fatigue at the same time meaning no individuals would be around to do the essential egg and larvae cleaning. They might survive the short term catastrophe better, but only at the risk of long term colony persistence. In contrast if there was a pool of workers with high thresholds for work they would not respond to the immediate catastrophe but when all the hard workers were exhausted they would be available to respond to perform the long term critical functions. So such a system would have a better long term survival and be favoured by natural selection.

To model this scenario the researchers considered a lattice consisting of 2500 cells with the lattice populated by a colony of 75 workers. Workers move from cell to cell in the lattice at random until they encounter a task. Tasks appear in the lattice at a fixed rate and have a variable difficulty to perform. If the task is above an individuals threshold they do it but then they are fatigued and they need to rest for a while until they recover. In the models they allowed the task appearance rate and the fatigue recovery times to vary, but task difficulty was fixed at a low level. The main model they were interested in considered two scenarios - one where all the individuals have a task threshold of 5 and one where the mean is 5 but some have much higher thresholds and some much lower (called the variable threshold situation). There were two measures of colony effectiveness. How many tasks they performed, and how long the colony lasted. Colonies went extinct when key tasks were needed to be performed but no workers were available to do them. When they ran the models the results were pretty much as they predicted from their alternative scenario described above. The variable system was always outperformed in the short term by the numbers of performed tasks because there were always workers doing nothing. However, long term persistence was higher because there were almost never times when everyone was fatigued, and vital tasks could always be performed. In the variable system there was always an individual with a high threshold who was able to do the colony critical tasks when the hard workers were all burned out coping with the more trivial stuff. The key to why high threshold workers are able to evolve is the presence of tasks that must be continuously performed to ensure colony persistence. If all workers have the same threshold for working then there is a strong chance that they will simultaneously all be tired and these critical tasks will not be performed. An unperformed critical task increases in importance (difficulty) until it is above the threshold of even a high threshold individual. Having a pool of high threshold individuals therefore makes sure that there is always someone able to step in and perform these critical functions.

The authors point out that their model is rather simple because it doesn't include the possibility that different workers preferentially perform different types of task. That is their thresholds are task specific. Such scenarios are found in many insect societies and of course are characteristic of human societies and organisations. They suggest however that even in such situations of individual task specialisation the fundamental observations that there are tasks that always need done, and individuals with low work thresholds will ultimately fatigue and cannot work all the time, will lead to selection favouring the development of a pool of high threshold individuals who are able to step in during critical times when the low threshold people are burned out.

Does this have any relevance for human societies and the world of work. Maybe it does. After all the scenario where the manager looks at his staff and wishes he could cut off the lazy tail are so common place that it seems possible that in human work we have the same situation. After all organisations flourish and fail in the same cutthroat type of environments that hone insect societies by natural selection. A workforce consisting of high and low threshold individuals may be more persistent because despite lower short term productivity they have greater endurance. Maybe the underperforming tail is a necessary feature of a persistent organisation. If you chopped it off then productivity might go up in the short term but only at the risk of greater persistence. It may seem counterintuitive but these models suggest that it isn't the firm with greater short term productivity that will last the longest. If you are a manager thinking about sacking your high threshold individuals then you might want to think again if the long term scenario is worth it. On a more domestic level if you have a teenager who is refusing to clean their room or do the dishes then perhaps one way to look at it is that they are not being lazy, but they are just have a high response threshold which makes them an important member of the community playing an essential role ensuring the persistence of society, (if not its immediate productivity!)

Shrinking birds

May

As a research student there is always something a bit special about your first international research conference. Here at last is a chance to escape the bench, and see in the flesh the people behind the names of those papers you have been reading. Perhaps, if you are lucky, an opportunity to capture them after a presentation in conversation, and tell them a little about your own work. Plus the frisson of nervousness and excitement that goes with giving your own first big presentation, outside of your own institute, and discussing it with like minded strangers afterwards. It is now 30 years ago but I recall vividly my own first conference like it happened yesterday. It was in the Netherlands, and was held at a field station operated by the University of Groningen, on the small island of Schiermonikoog. The trip was not only the first international conference I had attended, but was also the first time I had ever flown in an aeroplane, and only the second time I had gone abroad from my home country. So there were lots of reasons to be excited. I was particularly looking forwards to meeting two scientists who worked at the University of Groningen – Serge Daan and Rudi Drent, who in 1980 had published a landmark paper called 'The prudent parent', which addressed the question of what limits the level of energy expenditure, and hence what controls reproductive output. Since 1980 'The prudent parent' paper has been cited more than 1500 times, and the question it addressed, and their suggested answers, have since stimulated a large swath of my own research activity. By 1986, as I nervously climbed on board the aeroplane that would take me to Amsterdam, I already knew the paper almost off by heart. Professors Drent and Daan had organised the workshop meeting on Schiermonikoog mostly to be attended by newly qualified PhD students working on energetics in various animals, and I was lucky enough, on the recommendation of my PhD supervisor, to have been invited to attend.

I arrived a couple of days early for the meeting, and as we waited for the other students from around Europe to arrive, I slept on the couch of a PhD student from the University. His name was Marcel Kersten, and he was also going to the conference on Schiermonikoog. During the few days I stayed at Marcel's flat I learned a little of the Dutch language, some of which remains with me to this day – principally 'dree biere' which means 'three beers'. The reason I know how to say 3 beers, and not 2 beers or 1 beer, which would actually be considerably more useful, was because Marcel and I spent the whole time in the company of an undergraduate student from Groningen who was also called Marcel – Marcel Klaasen, and who would also be going to the conference. Perhaps we hit it off together because for our research work at the time we had all been working on the same topic: the energy balance of shorebirds.

Shorebirds are a fascinating group of animals that exist virtually unknown to most people in the general public. In the summer many of them breed in enormous numbers across the vast tracts of treeless arctic tundra across northern Canada and Russia. It is a remote habitat, with an extremely low human population density, of mostly nomadic reindeer herders, and it is rarely visited by outsiders. In summer the habitat is packed with the insect life that the shorebirds feed on, and there is 24h daylight giving them perpetual opportunity to exploit this food resource. But attractive as it may be in summer, in winter it is thrown into perpetual darkness and temperatures that can fall as low as -50 ºC. So the shorebirds leave in their millions, and fly down the coasts of the main continental land masses, to wintering grounds which stretch from the temperate zone right down into the tropics. But to most people this colossal migration is invisible. You won't see any of these birds in your local park or back garden, because they live, as their name suggests, on the sea shores. But even there, they are rarely seen, preferring mostly to live on the expansive mud flats that are found in large estuary areas, where they feed on small shrimps, snails and bivalve molluscs, like cockles. If you ever went to the sea shore and saw some small birds running around in front of the waves as they crashed onto the beach, almost getting drowned with each advancing wave, but always managing to outrun the water – then these are one type of shorebirds. So myself and the two Marcel's spent a wonderful 3 days sitting in open air cafes, having nerdy conversations about shorebird energy balance, while drinking repeated orders of 'dree biers'.

After my PhD I stopped working on shorebirds, but both the Marcel's continued. I have over the years since the meeting in Schiermonikoog crossed paths with Marcel Klaasen every few years or so at conferences around the world. He is now a Professor at Deakin University in Melbourne, Australia, where his interests have broadened, but at their core he still has a strong focus on migration of shorebirds. Two weeks ago, when my online issue of *Science* arrived in my email inbox, I clicked on it to open it, to see the front cover had a picture of a shorebird on it. A bird that in English is called a 'knot' to be exact. A similar picture is on the front page of this book. The knot is one of the most abundant and invisible of the vast army of shorebirds that migrates annually to the tundra and back. It has an interesting Latin name which is *Calidris canutus*. The species part of this name *'canutus'* is named after an ancient king who ruled the area that is now Britain, Denmark and Norway, and who was called Canute. Canute is famous because of a story involving tides that is supposed to have happened about a thousand years ago, in 1028. Canute was surrounded by a group of obsequious courtiers. These flattering courtiers told Canute that because he was king he had supernatural powers. To prove them wrong, and that only God could command the elements, he had them bring his throne to the shore when the tide was out. As the tide came in he commanded the waves to stop advancing – but of course they didn't, and he ended up with the water up to his knees, before the courtiers rushed in and rescued him. He then turned to his courtiers and said 'Let all men know how empty and worthless is the power of kings'. The knot is called *canutus* in Latin because they have the habit, when the tide is coming over a mudflat, of staying until the last possible moment, when the water is up to their knees, before they fly off to find a roost somewhere to wait for the tide to go out again, so they can resume feeding. In more recent years the knot has sometimes been referred to as the 'moon' bird. The reason for this name is because of an individual knot that was tracked in North America migrating between Canada and Chile and back, over a period of 18 years. The bird (called B95) flew over half a million kilometres during this period, which is equivalent to the distance from earth to the moon (and halfway back!).

Anyway to my delight it turned out that the knot was on the front cover of *Science* because inside there was a paper by my old friend Marcel Klaasen. The paper is about the complex consequences of climate change on this small shorebird. In fact in 2015 the population of knot that inhabits the north American tundra was already listed as 'threatened' by the US fish and wildlife service because of the decline in horseshoe crab eggs in Delaware bay on the east coast of the USA, which is a key stopover site on the long migration route south between Canada and Chile. The US population has already declined in some areas by over 75% since the 1980s. The paper by Marcel, however, concerned the population of knot that nests on the Russian tundra, and then migrates to the west coast of Africa to spend the winter.

Photo: Knot in winter plumage eating a small bivalve mollusc.

(photo: Jan van de Kam)

By analysing satellite imagery of the tundra collected over a 30 year period, they were able to show that the date when the snow melts has been getting earlier and earlier.

In fact the snow now melts about 2 weeks earlier than it did when we were sitting in the bars in Groningen drinking beer all those years ago. This has some important consequences. The plants revealed when the snow melts now start to grow earlier, and the populations of insects that feed on the plants emerges earlier as well so that the peak insect abundance occurs correspondingly couple of weeks earlier in the summer. This is a problem because the birds sitting on the beach in Africa are blissfully ignorant that the melt is getting earlier.

To time their flights north the knot have to guess when the snow will melt, and they do this probably using the cues from the daylength. So the time they fly north and arrive on the breeding grounds has not changed much, since the daylength is the same as it has always been. But the warming climate means that they now mismatch their breeding effort relative to the abundance of insects. The direct consequence of this mismatch is that the birds born nowadays on the tundra have less food available to support their growth, and consequently they are substantially smaller at maturity, and have shorter beaks than the birds that were born only 30 years ago. Shrinking size has been reported in many other species as a response to increasing global temperatures, and it has often been interpreted that this is advantageous because as it gets warmer a smaller bird, that has a greater surface to volume ratio, can more easily dissipate heat. So maybe the shrinking size of the knot as it gets warmer is advantageous. Unfortunately it turns out in this instance not to be the case.

Because the research team were able to study knot, not only on their arctic breeding grounds but also in Africa where they spend the winter, they could see the full extent of the consequences of their shrinking body size – and the news is not good. As I noted above in the winter the birds feed on large mud flats where their main food is small bivalves. The principal food in Africa is a bivalve called *Loripes*. The *Loripes* bivalves live buried in the mud, on average about 35 mm down, so the knot have to probe for them with their beaks. Unfortunately, if you have a shorter beak then you just cant reach deep enough to find them. So the shorter beaked birds have to resort to eating other things, like another bivalve called *Dosinia* which isn't buried as deep, but provides less energy per shell eaten and is less abundant, or sea grass (Zostera), which is just nowhere near as good as the bivalves, nutritionally speaking. The researchers showed that the proportion of *Loripes* in the diet was directly related to beak length, with birds that had beaks 30mm long having only about 19% of their diet as *Loripes*, while birds with beaks 41mm long included about 41% of their diet as *Loripes*. It is like if you had to use a stick to retrieve your lunch every day, but all the best food was further away than the length of your stick. This difference in diet then has a major impact on survival which is about halved in the shorter beaked individuals compared to the longer beaked ones. In theory that differential survival should lead to evolution of longer beaked birds, but as long as the birds time their journeys north too late to grow such beaks then such evolution isn't going to happen. Climate change it seems is just happening too fast for these birds to evolve a response, and the result is population declines on a global scale. I emailed Marcel immediately to congratulate him on his paper, but he was sanguine. "Thanks" he said "it's just a shame that such publication success has to come from such a sad story."

Women and children first, or last?

June

Picture the scene. You are standing by the side of a railway track. To your right the track stretches out to the distance and to your left it enters a small gorge with steep sides. In the gorge there are five workmen on the railway line. You look to your right and in the distance you notice a train is coming. The men in the gorge can't see it because of the gorge walls. It is clear that by the time they see it they will be unable to get off the track, and by the time the train sees them it will be too late for it to stop. A disaster seems inevitable. But then you notice in front of you there is a lever which would allow you to divert the train onto a disused side track. Problem solved. You can divert the train and avert the massacre. But then you notice something. You look down the disused side track and there is a lone worker sleeping on it. Both sets of workers are too far away for you to shout to them. If you run over to the single guy to get him to move you won't get back to the points in time to divert the train. What would you do? Would you switch the points and kill one person in order to save the other five?

This dilemma, often called the 'trolley dilemma' because in some versions it is a runaway tram trolley that is coming along the tracks, was developed in the 1960's by the philosopher Phillipa Foot. In the original version you are the driver of the train or tram and the brakes have failed. The people in this version are tied up on the tracks so they cant escape. You can see the two options in front of you. Would you steer onto the side track? The interesting thing about this dilemma is that humans immediately get the point of it. If you simply lay out the options to them, then most people realise that 'aim' is to minimise the loss of life. They do not for example see it as a competition where the objective is to kill as many people as possible. Faced with the trolley dilemma most people (90% on average) opt to switch the course of the train onto the side track thereby killing the single person and saving five. The trolley problem has been widely used as a tool to probe aspects of human morality.

In the 1970s the trolley dilemma was slightly reframed by Judith Jarvis Thomson, a moral philosopher from the USA. In Thomson's version of the dilemma you are not stood by a set of points, able to divert the train to kill a lone person on the side track, but instead you are on a bridge over the main track. Sitting on the bridge there is a fat man. If you push the fat man off the bridge into the path of the oncoming train you will stop the train and save the other 5 people. Would you push him off the bridge? When faced with this second scenario many more people say they would not. The two scenarios have the same outcome, but they produce radically different answers. There has been sustained debate for many years about why people say they would switch the points in the first scenario, but not push the fat man off the bridge in the second. It seems there is a large moral difference between 'doing' and 'allowing'. In the first scenario we 'allow' the train to run over the single person by changing the points. In the second we actually have to 'do' the act of killing ourselves, which seems to be more morally repellent.

An interesting practical question that has recently emerged from the trolley dilemma is to now imagine that you are the programmer of software for a driverless car! The car is going down the road and suddenly encounters a group of 5 people in the middle of the road. It is too late to stop and the people will be killed unless the car diverts its course. However, there is a single pedestrian standing on the sidewalk. Should you program the car in this situation to divert away from the 5 people and kill the pedestrian on the sidewalk? Should the car behave like a human responding to the trolley problem. Logically following the decisions of 90% of the population you should program the car to kill the single person on the sidewalk. But now consider this. You have programmed the car to behave in this way and it encounters the following situation. There is a pedestrian crossing and the light clearly indicates that pedestrians should not cross. A group of five drunken people come along. They see the red light but they ignore it and decide to cross anyway. A sober citizen is standing at the side waiting for the light to change before crossing. At that point the driverless car you programmed comes along and kills the law abiding citizen patiently waiting for the light to change to green, saving the five drunkards who ignored the red light. Was the programming correct? Should you, the programmer, be charged with murder?

The reason I am telling you about the trolley dilemma is that this month (June) there has been an interesting paper published which uses the trolley dilemma as part of its approach. The paper is published in the journal *Social Psychological and Personality Science* and it concerns the idea of chivalry. Chivalry is a code of conduct which has its roots in mediaeval Europe about 800 to 900 years ago. The chivalric code has many features relating to how one should conduct ones behaviour, which includes the ideas of honour in battle, and upholding the ideas of truth and honesty. Perhaps the most recognisable aspect of chivalry is the conduct that a man should have towards a woman.

In particular the idea that it is a mans role to defend a woman and to behave in a gracious manner to first his partner and then all other women. Perhaps this sort of behaviour is exemplified in no better way than the idea that if there is a disaster then it is the women and children who should be saved first. 'Women and children' first is most often associated with maritime disasters. The reason for this is that in the 1800s and early 1900s there were few regulations regarding the numbers of lifeboats that ships had to carry. So generally ships had inadequate lifeboat capacity to deal with the number of passengers and crew. The 'women and children' first rule was a legacy of chivalric behaviour in that it dictates that in this event it is women and children who should get the lifeboat seats, and the men should be left to die on the sinking ship. The first record of this actually being put into place was when the British ship 'The Birkenhead' sunk in 1852. The most spectacular and famous example, however, is the sinking of the Titanic in 1912. When it became obvious the Titanic was going to sink the captain was asked if the officers should start to get the women and children into the lifeboats. He replied "Yes, women and children in, and lower away". This seems to have been a costly statement because some interpreted it as 'women and children first' but others interpreted it as 'women and children *only*' and accordingly lowered half full lifeboats if there were no other women and children around to get on, refusing to allow men to board. Some men trying to board the half empty lifeboats were shot by the crew for trying to do so. When the final tally was counted 74% of the women on the Titanic were saved, but 80% of the men died. The harsh treatment the surviving men received from the press and society, that branded them as cowards, must have made some of them wonder if it would have not been better to have gone down with the ship.

In modern society where the feminist movement has made enormous strides for equality for women does this type of chivalrous code still guide our behaviour? The study which was published in *Social Psychological and Personality Science* was a collaboration between Columbia University in New York and Cambridge University in the UK, suggests that surprisingly it seems it does. In the second version of the trolley dilemma that I described above, for example, both men and women were far less likely to throw the person off the bridge if she was a woman than if it was a man. In the second experiment the subjects were given a sum of money (£20). They were told that the amount of money they had left at the end of a series of experiments would be multiplied 10 fold – giving them up to £200 as a reward. But there was a catch. Part of the experiment involved interacting with a partner. The catch was that the partner would be given an electric shock but the person holding the money could reduce the severity of that shock by paying out some of their money. Participants were much less likely to donate money to reduce the severity of the shock if their partner was male. In fact on average if the partner was female the participants ended up with only £8.76, while if their partner was male they ended the game with £12.54. These experiments suggest that there is a strong gender bias in our attitudes which harks back to the days of the chivalrous code of conduct. "Society perceives harming women as more morally unacceptable" said co-author of the paper Dr Mobbs from Columbia University.

But do we really. Are women really protected from being harmed relative to men? A problem with such thought games is that they do not actually involve us having to really do anything, and subjects know they are not real. No trains hurtle down sidings killing sleeping workers, nobody is actually pushed off a bridge to their certain death, and people playing the games know this. In reality our behaviour may be governed by much more pragmatic factors. Take the person on the bridge situation. If you pushed a person off a bridge in front of a train you would most certainly be charged with murder, and spend a great deal of time imprisoned or even be sentenced to death yourself. It would not be a valid defence to say that you were trying to save 5 people further down the track, because the prosecution would argue that you could not know that these people were certainly going to die. Hence you definitely killed one person for an unknown probability of saving others. In real life nobody is going to take these chances. Plus the trolley problem itself is so fantastically contrived that it may provide little insight into what people would do in real life moral dilemmas. There are in fact much more realistic versions of the generic trolley problem that result in dramatically different outcomes. The best known was also devised by Judith Jarvis Thomson, and involves the situation in a hospital where you have 2 people who need kidney transplants, and one person each needing heart, lung and liver transplants, all five of which will die otherwise. Also in the hospital is a guy with a broken leg. Should you kill him and harvest his organs to save the other five. Universally when people are presented with this scenario they do not opt to kill the guy with the broken leg to save the other 5, when in the morally equivalent trolley dilemma situation 90% opt to kill the lone worker instead of the other five. Perhaps it is the realism of what the consequences would be of the alternative courses of action that changes the outcome.

But what about the 'women and children first' phenomenon – we have real life data for that from, for example, the Titanic disaster, that people really do put women and children first. Or do they? Three years after the Titanic sunk, after hitting an iceberg, a British ocean liner called the Lusitania was sunk by a torpedo from a German submarine. One thousand two hundred people died, but the 750+ people who survived were dominated by fit young men aged between 16 and 35. Which of these two tragedies is the more representative of human behaviour? Four years ago a couple of Professors from Uppsala University in Sweden, Mikael Elinder and Oscar Erixson, set out to answer that question by analysing the data from 18 major maritime disasters over the last 300 years involving the deaths of over 15,000 people. Their work published in the Proceedings of the National Academy of Sciences of the USA in 2012 have some rather sobering findings that destroy completely any notion of chivalrous behaviour. When it comes to sinking ships, women have a distinct *disadvantage* compared to men, and this disadvantage seems independent of things like how long it took the ship to sink, which is frequently suggested to be the main difference causing the patterns of survival of people on the Titanic and the Lusitania.

Feminism has perhaps given women greater opportunities in general society, but when it comes to shipwrecks a woman's survival chances have actually got worse over the last 100 years. We may like the notions of chivalrous behaviour, protecting women, and the moral code of 'women and children first', especially when we play psychological games: but when the boat is really going down the rule people actually follow is "every man for himself".

Why do we not feel hungry when we are ill?

July

When I was 10 years old I was walking home from school one day and I got a really bad pain in my abdomen. It was excruciating and so bad that I had to crawl home the last 100 or so metres to our house on my hands and knees. When I eventually got home my mother was very unimpressed by the fact that my school clothes were all dirty from crawling along the path to our house. However, despite her shouting at me about it, all I could think about was the pain that was so severe I couldn't walk. I crawled onto the sofa and lay there holding my stomach and moaning. My mother initially ignored me because she was so annoyed, but eventually it was obvious that there was something seriously wrong. She called the doctor and he came to examine me. The next thing I knew I was in an ambulance and rushed to the hospital for emergency surgery to remove my appendix. Many years later my mum confessed to me that she felt really guilty about shouting at me for getting my school clothes dirty when I was so ill, but to be honest I couldn't remember her shouting at me anyway. In fact I don't recall very much about the whole episode, but one thing I do remember really distinctively was that as I recovered in hospital I almost completely lost any desire to eat anything. Meals would come and go untouched. Nurses would try to get me to eat. My parents would encourage me to at least try something. But I just didn't want to eat anything at all for at least 3-4 days.

In fact this response of appetite loss during severe illness is not unusual. It isn't just a response to surgery either. In fact it is a characteristic response to many types of infection, notably bacterial infections. All you want to do in these situations is find a quiet corner somewhere to lie down and recover. It is possible that this inhibition of appetite during illness is an evolutionary adaptation. There are two ways that this has been hypothesised to be potentially beneficial. The first idea is that perhaps the reduced nutrient intake when you stop feeding may reduce the supply of essential nutrients that permit the infectious organisms in your body to thrive. By starving yourself you are also starving them, and that might improve your speed of recovery. Another potential mechanism, however, is that if your appetite wasn't inhibited then you would have a strong drive to go out feeding. If you were ill, or had a broken leg, then you would not be performing at your best. So your chances of actually finding food might be much reduced and you would have wasted energy trying to find it. Not only that, you might be prone to not only failing to find your own dinner, but becoming the dinner of a predator yourself. So switching off your appetite circuits when you are ill may make a lot of evolutionary sense.

However, although your appetite has gone, your energy expenditure isn't switched down to the same extent. You are less active, and so you don't have to spend energy moving about, but you cannot switch off your resting metabolism – the energy that you spend all the time, whether you are exercising or not. Indeed your resting metabolic rate may actually be increased as you boost up your immune system to fight off the disease or repair damaged tissue. Individuals who have had severe burns for example have a profound increase in their resting metabolic rate that may last for up to 2 years following the event where they got burned. This mismatch between the rate at which food energy is being acquired via feeding and the rate at which it is being expended means we need to draw on our energy reserves to meet the energy demands. There are two main sources of such energy. The first is glycogen that is stored primarily in our livers. Typically the amount of glycogen we store is enough to keep us going for about one or two days without any food intake. After that we have to start drawing on our stored body fat. This is potentially why fat evolved in the first place – as a temporary storage organ for energy that we could use when we are ill. Interestingly people who are more obese tend to have better prognosis when it comes to serious illnesses like recovering from stroke, heart attacks and surviving chronic kidney disease. That is once you have a heart attack or a stroke you are better off being fatter than thinner. However, before you rush off to gain some weight it is worth noting that your risk of having a heart attack or a stroke in the first place is higher the fatter you are. This increased risk of developing such disorders is the main reason why obesity is regarded as a medical condition that requires treatment, rather than something good that will protect you if you have an unrelated medical issue.

Switching off hunger during an infection may be a useful adaptation for things like infections which are short-lived, but during much more prolonged illnesses like cancer or AIDS, turning off hunger can be a major problem. In these cases prolonged appetite loss may develop into a condition known as cachexia. All weight loss is generally a mix of energy taken from fat and energy from lean tissue. During cachexia individuals utilise a disproportionate amount of lean tissue. Since this includes muscle loss they rapidly become fatigued and in severe cases people become unable to perform even routine tasks. This is important because cancer and AIDS patients that develop cachexia have a much poorer outcome probability than those that are able to avoid it. In fact cachexia is said to be the primary cause of death in about 1 in 5 patients that has cancer, amounting globally to around 7.4 million deaths per year. At present about 50-85% of cancer patients develop weight loss before they are diagnosed. Indeed unintended weight loss is often one of the first things that people consult their doctors about prior to a cancer diagnosis.

Understanding the mechanism by which we lose our appetite during illness has therefore generated intense interest among scientists for two basic reasons. First, if we knew the mechanism we could perhaps intervene to provide more effective nutritional support for people who have chronic illnesses. One key aspect of cachexia is that it has a very poor response to simply trying to encourage people to eat more food (about the same level of success as trying to persuade obese people to eat less food). If we knew how the loss of appetite, and the preferential use of lean tissue, was signalled then it might be possible to intervene in the process to get people with illness related appetite loss to eat more. This might stem the progression into full blown cachexia. Plus of course while the key aim may be to get one group of people to eat more, there is a desperate need for drugs that encourage other people to eat less, so that they can avoid, or reverse, the problem of obesity.

Work on appetite loss and cachexia has been going on for a long time. Previous work published over 30 years ago had shown that during illness we produce a lot of compounds called cytokines which mediate the immune response to infection. These include members of the interleukin (IL) family (notably IL-1 and IL-6) and Interferon gamma. It has been long known that several of these cytokines can also have impacts on food intake, which was shown by injecting them directly into animals that were otherwise not ill to see what their food intake responses were. Initially it was considered that the most important mediator of appetite loss and the progression to cachexia was a cytokine called Tumour Necrosis Factor alpha (TNFa). However, during the 1990s it became clear that TNFa may not act alone but be complemented by other cytokines that work together in concert. In the early 2000's an additional player in this cytokine repertoire that interferes with our appetite was identified called interleukin-18 (IL-18). Mice that lack IL-18 eat more food, have reduced energy expenditure and eventually develop obesity.

How IL-18 works in the brain however was a mystery because there are no receptors for IL-18 in the main brain region that controls our food intake (the hypothalamus). Now a paper just published in the Journal of Neuroscience (May 2016) claims to have discovered a mechanism which suggests that IL-18 may be the key player that produces illness induced appetite loss. The work was performed by researchers based at the Scripps Institute located in California, USA. The jumping off point for the work was the finding that there are receptors for IL-18 in a small area of the brain known as the Bed nucleus of the Stria Terminalis (or BST) which is part of the amygdala. The BST is a really interesting area of the brain for reasons completely unrelated to control of appetite. Apparently the size of the area varies enormously between men and women. On average males have a BST about twice the size of that in females. Interestingly a study of six male to female transsexual subjects had female sized BSTs. This was apparently not related to the treatment of such subjects with the female sex hormone oestrogen, because a single untreated transsexual had a similar female sized BST. In contrast a female to male transsexual subject had the typical male number of neurons in the BST. Hence the BST may be a key area of the brain involved in gender identity disorders. Strangely abnormally small BSTs have also been reported in convicted male paedophiles.

A subset of neurons in the BST send projections to the hypothalamus. These neurons were the ones carrying the IL-18 receptors. When the researchers injected IL-18 directly into this area of the brain in mice they found a significant reduction in food intake that lasted for at least 6 hours. Mice injected with IL-18 ate only 0.7g over the 6h compared with 1.4g in the controls. Injecting mice with IL-18 had no impact however on their physical activity levels, or their levels of energy expenditure, showing that the effect was only related to the intake side of the energy balance equation. The researchers took slices of mouse brains through the BST and kept them alive in saline while they recorded from various cells using a technique called patch clamping. This method involves attaching an electrode to individual neurons and recording its electrical activity. When the researchers monitored the cells from the BST they found that a neurotransmitter (glutamate) strongly activated the neurons that project to the hypothalamus – called the type III neurons. This excitatory input to the BST cells was strongly reduced when they were treated with IL-18. Because these neurons project to the hypothalamus, reducing the excitatory input to the neurons had a knock on effect on the activity of neurons in this region of the brain. In fact, treating cells in the BST with IL-18 increased the firing of neurons in the hypothalamus, because the cells coming from the BST had reduced output of another neurotransmitter (GABA) which normally inhibits the hypothalamic neurons. The pattern of changes recorded in the hypothalamus was consistent with the patterns expected when animals experience a reduction in appetite. To show that this pattern was really due to IL-18 the authors constructed a mouse in which the IL-18 receptors were absent in the brain. As predicted, if IL-18 was causing the effects, when brain slices from these mice were treated with IL-18 there was no effect on the firing pattern in the hypothalamus.

Identifying IL-18 as a key component of the reduced appetite that accompanies illness opens up several opportunities to intervene in the process. For example, if the IL-18 receptors in the brain could be treated with a drug that blocked the action of IL-18, then elevated IL-18 during illness would no longer suppress appetite. The main question that remains unanswered is the extent to which IL-18 mediates the illness effect alone or in concert with other cytokines. The critical experiments where mice lacking the IL-18 receptors were given an infection to see if the infection related reduction in appetite was reduced were not performed in the paper. On the other hand, even if it proves to be the case that reversing illness induced appetite is only partially successful, using other drugs to stimulate these receptors in the BST might inhibit food intake, and hence be a novel treatment option for obesity. This seems a more likely outcome because it was clearly demonstrated that injecting IL-18 itself into the BST region of the brain reduced the food intake of mice by over 50%. The problem as always with putting drugs into the brain will be potential side effects – and given the wider role of the BST in sexual identity and paedophilia, these could be, to say the least, interesting.

Can we make our fat burn itself?

August

Our bodies have three types of fat. The first, and one that most people will be familiar with, is white fat (or white adipose tissue: WAT for short). White fat acts as an energy reserve to store energy when our energy consumption as food exceeds our expenditure, and it is a vital reserve to draw on when our energy intake falls short of our demands. White fat cells have a single large droplet of fat in them surrounded by a thin layer of cytoplasm (see figure). Obesity is an excessive storage of white fat, and it is a problem because it leads to all sorts of health problems like type 2 diabetes, cardiovascular disease and cancer. The second type of fat is brown fat (brown adipose tissue or BAT) (see figure). Brown fat cells look very different from white fat cells in that they have fat droplets in them, but there are generally several droplets instead of one large one. In terms of volume, there is a much greater ratio of cytoplasm to fat droplets in BAT, and the cytoplasm is filled with mitochondria. Because of their lower fat content the cells have a brownish colour and hence the name. Mitochondria are often called the power houses of the cell because they are where the diverse micronutrients we consume as food are converted into a currency that can be used as a fuel source to power all cellular functions called ATP. The mitochondria in BAT cells however have a special protein that actually undercuts their ability to manufacture ATP. As a result when this protein is activated the cells release the energy, that would have gone to making ATP, directly as heat. The protein is called uncoupling protein 1 because it uncouples the movement of protons in the mitochondria from the production of ATP. The main function of BAT is to generate heat for thermoregulation. BAT is sometimes called the 'internal furnace' because of this role. BAT is really abundant in the bodies of small mammals which need lots of heat for thermoregulation because they have a large surface to volume ratio. For a long time it was thought that, although humans have BAT when we are babies, it disappears when we are adults. However in 2007 it was shown that actually adult humans also have functional deposits of BAT.

About 25-30 years ago several groups noticed that when small mammals like mice are placed in the cold their WAT stores start to take on a browner appearance – a process now called 'browning'. There was some debate over what was actually happening in this process with some claiming that the brown cells were a result of the cold stimulating

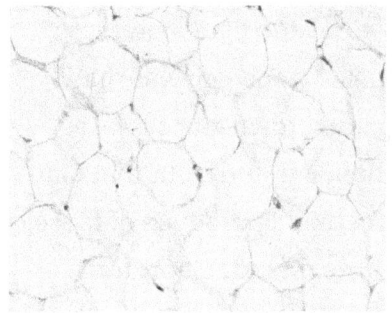 White Adipose tissue: note the cells are filled with a single large fat droplet.

Brown adipose tissue; note the cells have lots of small fat droplets in them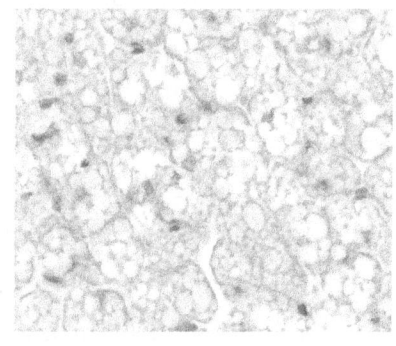

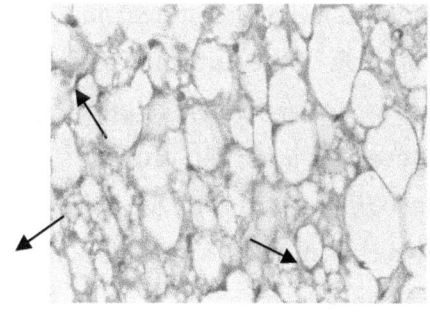 White adipose tissue after cold exposure: the tissue consists of a mix of white and brown cells (or Brite cells)

development of new brown cells within the white adipose

tissue. One scientist from Italy, Saverio Cinti, however suggested that what was actually happening was that white fat cells were changing into brown fat cells by a process called trans-differentiation. It turns out that Cinti was partially correct. In fact there is a third type of fat cell that have become known as brite fat cells (short for **BR**own in wh**ITE**) or 'beige' fat cells again reflecting their colour (see figure). These brite/beige cells actually come from a separate lineage of cells that are separate from true white fat cells. When they mature they can display either a white or a brown appearance as the needs change between energy storage and heat production. The discovery of brown/beige cells has created an enormous amount of work in the obesity field to try and dissect the molecular mechanisms that underpin the interconversion. The basic reason is that the function of the brite/beige cell in its brown phase is, like the true brown fat cell, to burn energy. So if we can switch these cells into their brown phase, by browning our adipose tissue, it may be feasible to burn off the energy stored in the true white fat cells – and the obesity problem would be solved.

This is a really attractive prospect because current pharmaceutical approaches to obesity have relied on two strategies that, in practice, are both somewhat flawed. The first tries to interfere with the hunger centres in the brain. The problem with this approach is that the hunger centres of the brain are tightly linked into other emotional centres. Several promising weight loss drugs have already been withdrawn because of side effects on depression and suicidal thoughts. The second pharmaceutical approach is to interfere with fat absorption in the alimentary tract. This also works but with the downside that the fat that is not absorbed has negative effects on the composition of the faeces and in exceptional circumstances can lead to 'rectal leakage', which as you may imagine is not popular among patients using these drugs. So switching on brown fat or converting brite/beige fat cells from their white to brown form is an appealing third option.

A key question, however, is whether it will actually work. Rather than focussing on the molecular biology we decided that a proof of principle test of this could be performed in a rather different way, and the paper describing these data was just published in the journal *Scientific reports*. Our rationale was as follows. Research has previously shown that human brown and brite/beige fat is activated by exposure to cold, very much as happens in small mammals. For example, it has been shown that the levels of brown and brite/beige fat in people living in both Japan and Europe are increased during the winter. So if increased levels of brown and brite/beige fat are found under colder conditions, and if these higher levels are able to burn off excess ingested calories then people in areas where it is colder should be thinner. This is not a completely new idea but previous attempts to investigate the relationship between temperature and obesity have been plagued by problems of confounding factors. For example, the countries with the leanest populations in the world are to be found in the extremely hot countries bordering the Sahara desert. Yet countries with some of the fattest populations are also found in the extremely hot countries of the middle-east – such as Kuwait for example. The main differences between these countries, apart from profound cultural differences, are their levels of wealth and poverty which are clearly more significant factors driving the obesity epidemic on this type of scale. To avoid these issues we decided to explore the link between obesity (and type 2 diabetes) prevalence and ambient temperature using only data from the United States of America. This avoids many of the issues of cultural differences between countries, but also the USA is an ideal testing ground because they monitor obesity levels carefully using annual surveys involving many thousands of subjects that are supported by the USA Center for disease control. This is combined with the fact there is also a large network of climate monitoring stations that permit detailed analysis of the ambient temperature levels across the US and the USA census which also provides a wealth of demographic data that can be used to avoid the impact of confounding factors.

This latter point is important because the USA, while more culturally homogenous than comparisons made globally, still has regional variations in two key factors that also drive obesity (and type 2 diabetes) levels. These factors are the racial make up of the population and the levels of poverty. Poverty is important because once a nation is developed it is actually the poorest people who show the highest levels of obesity. Moreover, certain races such has African Americans have a greater propensity to obesity even when poverty is accounted for, and they also more readily develop type 2 diabetes at any given level of obesity. The mainland US is divided into 49 states which are each divided into counties. Counties have variable populations that average about 64,000 individuals. In total we were able to recover data on levels of obesity, type 2 diabetes, poverty, race and temperature from 2654 of the 3146 counties in the mainland USA covering around 170 million people. When we just looked at the raw data there was as expected a positive relationship between the average ambient temperature in a county and the obesity prevalence. However the effect was small, with ambient temperature explaining just 5.7% of the variance. The average level of obesity in a county with an average temperature of 5 °C was 29.6% and that in a county with an average of 25 °C was 33.6% - i.e about 1.13 times higher in the warmer conditions. Surprisingly however the effect of ambient temperature on type 2 diabetes was much stronger – explaining 29.6% of the variation. For type 2 diabetes a county with an average temperature of 5 °C had a prevalence of 7.6% while one with an average annual temperature of 25 °C had a prevalence of 12.1% (1.59x higher).

A key question is whether these effects were just an artefact of the distribution of poverty and race, because warmer counties on average have higher populations of African Americans and are also on average poorer. In fact, when we normalised the data for these trends the effect of ambient temperature on obesity completely disappeared. But, although the effect was reduced, there was still a very strong impact of ambient temperature on the prevalence of type 2 diabetes. This relationship still explained 12.5% of the variance in diabetes prevalence. Perhaps most interesting the strength of this relationship depended on which month of the year the temperatures were recorded. During July, when everywhere was quite hot (range 20 to 30 °C) the variance explained was only 3.3%

while in January, when there were much large differences between the counties (range -15 to +15 °C) it was 16.8%. To set these figures into some sort of context, 16.8% of the variation in diabetes exceeds variance explained by the combined effects of all the genetic polymorphisms that have been linked to type 2 diabetes by genome wide association studies, which stands at around 10%.

One interpretation of these data is that in colder conditions our brown and brite/beige fat cells get switched on. It is known that these internal furnaces are extremely efficient at disposing of ingested sugars and fats and this may then be protective against development of type 2 diabetes. However, there are other possibilities. Colder areas are on average further north and hence colder counties also have different photoperiod regimes. Photoperiod may in turn affect vitamin D status and vitamin D status has been linked to the risk of developing type 2 diabetes. This is an unlikely explanation however because lower vitamin D levels in areas further north would be expected to lead to greater levels of type 2 diabetes, yet it was in these colder more northerly regions that diabetes levels were the lowest. Daylength may also be linked to depression and seasonal affective disorder, and these too have been linked to greater risks of developing type 2 diabetes. Again the direction of this potentially confounding factor is the opposite of the expectation. Finally, if we included latitude into the predictive model, the impact of temperature actually increased, rather than becoming smaller, suggesting that colder temperatures were not having a beneficial effect simply because colder regions were more northerly. The message seems clear. Switching on your internal furnace of brown and brite/beige fat may not do very much for your obesity levels, but it may have a strong beneficial impact on type 2 diabetes.

At the end of the day such epidemiological modelling is interesting but it is ultimately only correlations and it may be that despite our best efforts to control for any confounding factors that might create a spurious relationship it is always possible that there was something we didn't control for and the link of temperature to type 2 diabetes is just an artefact. We were delighted therefore that as our paper was in review a small clinical trial was published in *Nature medicine* which showed enormous beneficial effects on insulin sensitivity of exposing type 2 diabetic patients to cold temperatures (15 °C) for 6 hours per day. The interesting thing was these benefits occurred despite only small changes in their brown and brite/beige adipose tissue. This suggests there is something special about being exposed to cold that may have causal impacts on type 2 diabetes risk that we don't yet fully understand.

If you close your eyes, how do you know where your legs are?

September

Last year I had an interesting experience. Interesting, but not one I want to repeat any time soon. It happened when I was in a meeting where different companies were bidding to sell us a piece of equipment. When we arrived at the building, where the meeting was held, I felt a bit strange. We were waiting to cross the road. I turned my head to look at the traffic, but it took a second for the world to catch up with my head movement. It was like the world had some sort of inertia in it. I thought it was odd, but not serious, and then when we arrived at the room where the bids were to happen it occurred again. I turned my head towards the door when I heard someone entering, and the rooms took a second to swirl around and catch up. I then started to feel seriously disoriented. It was like being completely drunk, but completely sober at the same time. I asked if it was possible to go and lie down somewhere, and fortunately the company had a small reception room with some comfortable chairs in it. However, when I lay down I felt even worse. I broke into a cold sweat, and my heart was racing. My PA called for the company paramedic who came and took my blood pressure and listened to my heart. "Is this his first heart attack" she asked my PA. I was pretty sure I hadn't had a heart attack, because by all accounts they include severe pain, and I wasn't in pain at all. However, I was equally sure that something wasn't right, because every time I tried to stand up I fell over. An ambulance was called and I was rushed to a local hospital. The trip itself was eventful, as the ambulance managed to crash into a parked car as we prepared to depart and set off its alarm. Plus racing round the fourth ring-road in Beijing in an ambulance with its sirens going and lights flashing is exciting. Particularly when you are convinced that what you have is not life threatening.

In the hospital it was confirmed that I hadn't had a heart attack, and they then started a series of tests to find out what was actually wrong. In one of them the doctor held a pencil up in front of my face, and moved it to the left and right asking me to track it with my eyes. You have probably done it at some point, as I had previously. But this time I couldn't do it. When he moved the pencil my eyes kind of jerked backwards and forwards trying to locate it. It was extremely weird. Then I got this very odd sensation that the left side of the bed was falling away and

I was in danger of falling out, and as I tried to compensate for it I just about fell out of the opposite side! Eventually I was diagnosed to have a viral infection of the semi-circular canals in my inner ear, which provide our sense of balance. There was no cure. I just had to wait for the infection to subside. I persuaded them to let me go home. When we got out of the taxi, several of my students helped me walk back to my apartment, as I still couldn't stand up. I am sure to our neighbours it looked like I had had one too many drinks. What followed was the worst 5 days. My balance was so bad I couldn't walk around so had to crawl everywhere. I couldn't watch the TV or use a computer. I couldn't read anything without feeling nauseous, and more than once when I tried to write something I was actually sick. So had to sit there with my head perfectly still facing forwards and do nothing. It was like being a corpse. I was so happy when it started to subside and I could start to live again.

I think for me one of the strangest aspects of this experience was that we take our sense of balance as something completely for granted. We never really think about it. It is obviously there all the time. But it is only when it becomes impaired that we actually realise how critical it is to our whole existence. When mine went wrong I was completely debilitated. Another sense that we take almost completely for granted is our sense of where we are. I don't mean like the sense of being in Beijing, or in the UK. I mean that if I close my eyes I know exactly where my arms and legs are located in space. If I lift up my left leg but not my right one I can sense where it is in space without looking. If someone was to passively move one of my limbs I would know where it had been moved to. This sense is called proprioception. It is sometimes called the 'sixth' sense.

Not everyone has a perfectly working proprioception system. For some people if you get them to close their eyes and you then passively move their limbs backwards or forwards they are unable to tell you which way their limbs have been moved. Such patients have been known for over 100 years since Charles Scott Sherrington wrote a paper in 1906 defining the term proprioception. Scientists are now starting to unravel the genetic mutations that may underlie these strange conditions, and a big advance in this area was published this month in the New England Journal of Medicine.

The paper concerned two patients who were discovered because they shared a number of proprioception deficits. They were both female, but one was 19 and the other only 9. The two first came to the attention of one of the authors of the paper Carsten Bonnemann who works at the NIH in Bethesda Maryland in the USA. Bonnemann noticed that the two shared several unusual symptoms. In particular they would routinely sit with their fingers, feet and hips held in unusual positions, and had unusual spontaneous movements of their arms and hands. They both had extreme difficulty walking and had an exceptional lack of co-ordination. When the subjects were blindfolded they found it almost completely impossible to walk. They would stumble and fall. But with the blindfold removed they could walk – if awkwardly, and in one case only with support. When they were growing up both of the patients had not managed to walk unaided until they were between 6 and 7 years old. Similarly, they both had ongoing problems self-feeding and self-dressing. Bonnemann put the patients through a series of tests. In one test they had to place a finger on their nose and then move it to a target in space about 50 cm in front of them. When they were not blindfolded they could do this task easily. But when they were blindfolded they performed very poorly on the task compared to normal people, who generally perform almost as well blindfolded as fully sighted, because they can remember where the target is and they can then control their arms to move in space to the remembered location. If you log on to the web site for the journal you can watch a video of the patients and control individuals performing the tasks, which also show the unusual angles in the joints of their fingers and their problems walking (http://www.nejm.org/doi/full/10.1056/NEJMoa1602812). The videos are frustrating to watch as you see the people repeatedly try to locate the target. The patients also both had unusual abnormalities in their sense of touch. For example, they might perceive something that was soft (like a soft brush) as prickly, or they may not even sense it at all. Interestingly in addition to these behavioural abnormalities they both showed an extreme curvature of the spine (called scoliolis).

To see if they had anything genetic in common, their genomes were screened and it was found that they both shared a mutation in a gene called *PIEZO2* which stands for Piezo-type mechanosensitive ion channel component 2. This gene encodes a large protein that sits across the membranes of cells and appears to act as a sensor of mechanical pressure. Mutations in this gene had already been implicated in the sense of touch and co-ordination and a set of disorders known as arthrogryphoses. One such disorder is known as Gordon syndrome and involves subjects also having permanent fixation of their fingers and elbow/wrist joints in unusual positions, just like these patients had.

So effects of mutations in PIEZO2 on proprioception are not exceptional. The unexpected thing about the two patients discovered by Bonnemann and colleagues was that the mutation they shared was a complete loss of function mutation. Another surprising aspect was that, although the two patients were unrelated to one another, they had the exact same mutation (among other mutations that were unique to each person). The common mutation was actually a premature stop codon, and this type of defect means that the person can't make a functional protein because the message to build the protein carries an instruction to stop half way through the sequence. They didn't have just a mutated gene, but in effect didn't have the gene at all. This was really unexpected because when the gene had been knocked out previously in mice it had been fatal. So it was assumed that it would not be possible for humans to live without functioning copies of the gene. However, last year a paper was published in *Nature neuroscience* where the gene was knocked out in specific mechano-sensing neurons, and these mice survived, but they had severe disruptions of their co-ordination and limb movements, very like the patients in the NEJM paper.

The abnormalities in the patients sensations of touch were further investigated by studying them in an MRI machine which could show areas of their brain that were active when they were touched by various objects. For most people being stroked by a soft brush is a pleasant sensation. For these two patients however this was perceived as prickly, and a negative sensation. Interestingly in the MRI it seemed that when they were stroked by the brush the area of the brain that in normal people detects physical sensation wasn't activated at all. However, what was activated was a different region that is linked to the emotional response to touch. It was like the subjects couldn't actually detect the brush was touching them, but instead they could just feel an emotional response to its touch. The fact the patients seemed unable to sense things directly was confirmed by holding a vibrating tuning fork against their skin, which they seemed unable to detect. Despite these major abnormalities in their sense of touch they were actually completely unimpaired when it came to other reactions linked to touch – for example they did not have any impairment in their ability to feel pain, or to react to changes in temperature of an object they were holding.

The effects of the mutation on the shared skeletal

deformities in the patients was also unexpected and leads to some interesting ideas. The authors speculate in the paper that the sense of proprioception may be essential for normal development of the skeleton. That is if the body can't sense where the skeleton is located it may be difficult to direct how it should develop. There are also wider possibilities. Everyone knows someone who is a bit clumsy. They seem to lack fine motor skills. Often spill their coffee and food etc. Maybe these people have mutations in the same gene just they are not as severe as in the two patients in the NEJM paper. And then there are people who have exceptional dexterity and fine motor control, like pianists, and other people who play complex musical instruments. Maybe they also have mutations that make their fine motor skills better. The difference between a concert pianist and you may not just be their musical ability. The role of this gene is just starting to emerge and may have widespread importance for our sense of where we are.

No more 'bad air'

October

Two thousand years ago the citizens of Rome, the epicentre of the Roman empire, were beset by fevers that were so pervasive they became known for a while as 'Roman fever'. Sufferers had a classical cycle of recurrence of symptoms that included a sudden feeling of coldness that was then followed by a severe fever. In addition to elevated body temperature, which would precipitate profuse sweating, the fevers would include headaches, shivering, pain in the joints, convulsions and vomiting. The symptoms would subside but then recur again on a 1-3 day cycle. Some people might have a quick cycle of just a day but in others it might recur at 2 or 3 day intervals. Complications would occasionally include seizures and coma, and a distinctive whitening of the retina. More routinely sufferers would show acute respiration problems and severe anaemia. The skin might become yellow, reflecting impaired liver function, and the urine would occasionally become black. Eventually death was not unusual, and could occur within hours or days of the first symptoms appearing. This was particularly so for the unborn foetuses of pregnant women who became infected. Even among those who recovered, the fever might return several months later, most usually after being absent over the winter period. It has been suggested that the disease may have contributed significantly to the decline and ultimate fall of the Roman empire.

It was noted that the fever was most common in people living near to the marshy areas around the city, where there were lots of areas of stagnant water. For this reason the disease also became known as 'marsh fever'. Rome was not the first place to become affected. Similar symptoms of periodic fever were documented by Hippocrates in ancient Greece four hundred years previously. He had also noted the characteristic pattern that some people has cycles that took 1 day but others took 2 or 3 days. Indeed, a medical manuscript from ancient China 4700 years ago documents symptoms that were also the same.

You may think that marsh fever is just a thing of the past. But you are wrong, it is still very much with us today. On the 9th of December 2015 the World Health Organisation published their latest report on marsh fever, which showed that in 2000 there had been over 262 million people infected with it, of which somewhere between

650,000 and 1.1 million had died. If you were to lay the coffins down end to end they would have stretched from Beijing to Shanghai. At that point it was the leading cause of death among 2-5 year old children in Africa. Although great progress has been made over last 15 years, in 2015 there were still nearly half a million deaths from Marsh Fever, mostly among children under 16. Half a million is still about one person every minute, every day, all year long. So why have you never heard of it? The fact is you almost certainly have – but not by its old name marsh fever. In the middle ages Italian medics renamed marsh fever after the stinking air that pervaded the marshes where it was predominantly caught – they called it 'bad air', or in ancient Italian 'Mal aria'.

We still call it malaria today, although we now know that it is not caused by bad air but by a protozoan parasite (called *Plasmodium*) that gets into the blood via the bite from an infected mosquito. The mosquito has a larval and a pupal stage both of which are aquatic and require stagnant water for their development. Hence the link between the disease and areas with marshland. The recurrent fevers that characterise malaria are caused by eruptions in the parasite numbers in the blood, with different species of *Plasmodium* taking different time periods to multiply, causing the different periods between the fevers. The fight against malaria has been long and hard and has been based on a two pronged strategy. The first prong is to kill the mosquitos that carry the parasite, and to prevent as far as possible the insects biting victims and infecting them. In the 1950s and 1960s this involved a large amount of spraying stagnant water with pesticides like DDT to kill the larvae and pupal stages in the mosquito life cycle. This was really effective but at a major cost to the environment, because the DDT didn't only kill the mosquitos but just about everything else. The main approach nowadays is therefore to focus on using insecticide treated nets that protect people while they are sleeping from being bitten by the adult mosquitos. This is a much more targeted strategy with minimal environmental collateral damage.

The second approach is to develop drugs that kill the protozoan itself. The first drugs were developed by the practitioners of Chinese traditional medicine around the same time that the Romans were suffering from the Roman fever 2000 years ago. In South America Peruvian Indians discovered that the bark of a tree was effective against malarial fevers. The tree was named *Cinchona* by the great Swedish naturalist Linnaeus after a French countess who had suffered from malaria and was cured by taking a remedy made of the tree bark. In the 1600's the bark was imported into Europe, by missionaries who had been in South America, where it became a popular malaria treatment. However, it took almost 200 years for the active ingredient to be discovered – an amine called quinine. Even then it was not economically viable to artificially manufacture it, and the main source remained the bark of the South American *Cinchona* tree. From the mid 1800s onwards quinine became the major treatment for malaria across large areas of the world. Quinine has a bitter rather unpleasant taste. During the period when the British empire included large parts of India, that were rife with malaria, British colonials used to drink quinine as a 'tonic'. To make the quinine tonic more palatable they used to mix it with the alcoholic drink gin. This is apparently the origin of the cocktail drink 'gin and tonic'. This probably also explains the origin of the saying that people often make when they are drinking gin, that are doing so just for 'medicinal' reasons. An interesting aspect of quinine is that it absorbs ultraviolet light at 350nm and re-emits it as an intense blue light at 460nm. If you shine UV light on a bottle of tonic water it consequently glows bright blue! As a treatment for malaria quinine was largely superseded in the 1920s by derivatives that were more effective like chloroquine.

A key issue with drug treatment is that the parasite develops resistance to the drug and hence there is a constant need to develop new drugs that can be used in areas where the parasite is resistant to chloroquine and quinine. In the 1960s a researcher into Chinese traditional medicine (Tu YouYou) started a systematic search of treatments for malarial fever, in a secret project called the 523 project, which had been championed by Premier Zhou Enlai. Her search through ancient Chinese medicine texts led her to plants of the genus Artemesia (qinghao) and ultimately to a class of anti-malarial drugs called artemisinins. These drugs have become the drugs of choice in the frontline battle against malaria. In 2006 it was recommended that quinine and derivatives should only be deployed when artemisinins were unavailable. For her work in developing the drugs Tu was awarded one of the premier prizes in medical science in 2011, called the Lasker prize. In their 2015 report the WHO estimated that the use of artemesinins between 2000 and 2015 has been responsible for preventing about 400 million cases of malaria, and avoiding around 2 million deaths. Little surprise then that in the same year Tu YouYou received the Nobel prize for medicine and physiology. She is the first Chinese scientist to have received the Nobel prize for work completed entirely within mainland China. Since winning the prize she has also been called the Professor of three no's: no Phd, no time spent working abroad and no membership of the Chinese national academy! I predict at least one of these is unlikely to remain so for very long.

The holy grail of malaria research, however, is not to use drugs but to develop a vaccine, and the research effort in this respect has been intense and prolonged, spanning at least the last 30 years. At the frontline of this battle is a US company called Sanaria, which was founded in 2003 by a scientist called Stephen Hoffman. The goal of Sanaria is to develop a vaccine particularly to treat children in Africa. The company name is a nice twist on the word malaria. Malaria means bad air. Sanaria means 'clean air'. In June this year Sanaria published a key paper in the journal *Nature medicine* which described the results of a trial into their vaccine PfSPZ. PfSPZ stands for *Plasmodium falciparum* sporozoite. *Plasmodium falciparum* is the main species of the protozoan that causes malaria related deaths in African children. The vaccine actually contains live but weakened, purified malaria parasites that do not cause illness. The trial included 101 adult volunteers, ages 21-45, who were divided into groups, with 57 participants receiving the vaccine at varying doses to evaluate alternative dosage regimens. The vaccine was well tolerated and safe. Vaccine efficacy among volunteers was tested at 3 weeks, 21–25 weeks, and 59 weeks after the final vaccination. At 21 weeks after last immunization, 9 of

14 volunteers (55%) were protected against exposure to the bite of malaria parasite-infected (disease transmitting) mosquitoes. Five of the 9 volunteers protected at 21 weeks underwent repeat exposure to malaria parasite-infected mosquitoes at 59 weeks and all five (100%) were protected against infection. The main breakthrough here was the duration of the protection. The basis of this protection appeared to be dependent on immune cells (called T cells) that were mainly sequestered into tissues (particularly the liver) from where they could respond to an infection when required. Hoffman, in a statement published after the paper was published said "We intend to use PfSPZ Vaccine for preventing malaria in individuals, halting transmission of the malaria parasite and eliminating the parasite through mass vaccine administration campaigns. It's reasonable to suggest that within three-to-four years a safe, reliable vaccine could be a commercial reality and provide medical benefit to a huge population." This may actually be a conservative estimate. Last month (September) Sanaria announced it had received U.S. Food and Drug Administration (FDA) Fast Track designation for the PfSPZ Vaccine. This is the only malaria vaccine to have ever received this distinction. The 'Fast Track' is a process designed to facilitate the development, and expedite the review of drugs to treat

serious conditions. The purpose is to get important new drugs to the patient earlier. A drug (or vaccine) that receives Fast Track designation is eligible for Accelerated Approval and Priority Review, which can considerably speed the time to market. Sanaria's goal is rapid deployment of the vaccine into the key areas of Africa where it will have maximal impact on the local populations, followed by more commercial uses in the military, diplomatic service and tourist industry. Soon there may be no more 'bad air'.

Smelling bee

November

There is a famous quote, often ascribed to either Charles Darwin or Albert Einstein, that if bees were to go extinct then humans would also go extinct within 4 years. The argument is that the bee is so vital to pollination that its disappearance would lead to the loss of around 100,000 plant species. This includes many plants that are staple foods, either directly eaten by humans (such as most of our commercial fruits), or plants eaten by the domestic animals that humans also eat. Consequently, the saying goes, no bees, no pollination, no plants, no animals and hence no man. The fact some people say that this was said by Einstein, and others say it was said by Darwin, leads to doubt that it was actually ever said by anyone. In fact, Darwin did write about the importance of pollination services of bees in his most famous work: "The Origin of Species". In a section about how life was an interconnected network of plants and animals, Darwin said that if bumble bees were to die out, then this would probably have severe consequences for two common British plants : heartsease and red clover, both of which are almost completely dependent on bees for their pollination. Darwin stopped short however of carrying this loss of plant life to further impacts. Others however were less cautious in their extrapolations. The British Bee journal in 1887 carried the implications much further, coming to the conclusion that the existence of domestic cattle, and hence the beef we consume, was actually dependent on cats! The argument was as follows. Red clover is dependent on bumble bees for its pollination. Bumble bees make their nests on the ground, where they are predated by mice. And mice are eaten by cats. Thus if the cats were lost, mouse populations would expand and devastate the bee population. Pollination of clover would stop. Without clover there would be no food for cattle, and without cattle there would be no beef. The argument stopped tantalisingly short of saying that without beef there would be no humans. But that detail appears to have been added just 14 years later in 1901 by the Belgian author (and later Nobel prize winner) Maurice Maeterlink in his book "The life of the bee". Maeterlink was also responsible for the idea that bee pollination would not only eliminate populations of red clover but populations of over 100,000 plant species. The number was never justified, but it stuck, and was repeated many times in various sources, being eventually ascribed, also without justification, to Darwin himself.

How Einstein became involved in the quote is a little less clear but one idea is that it stems from a letter that a group of school children sent to Einstein in 1951 asking what would happen if the sun was to burn out. He replied "No sun, no wheat, no bread, no grass, no cattle, no meat, no milk…no life". Although Einstein's chain of dependencies didn't include bees, it is similar to the dependency chain that was also circulating which did include bees, and it seems they may have become confused. By 1966 the time limit of 4 years had been added from an unknown source, but also attributed to Einstein.

Reality is of course more complex and probably less dramatic. It is unlikely that if bees became extinct that whole lineages of plants would also die out, and if we stopped eating beef, the whole of civilisation would not collapse. There are large populations in India that are already surviving very well without eating beef. Indeed, the suggestion we should stop eating meat has been advanced as a method for us to cope with the potential world population of 10 billion that demographers confidently predict we will need to feed by the latter part of this century. But if bees did disappear then it would definitely have some large impacts – even if the complete collapse of civilisation within 4 years isn't one of them.

The reason this quote is often in the news recently is that over the last decade the prospect that bee populations may die out has actually become a reality – at least for populations of the western honey bee (*Apis mellifera*) which is the dominant bee responsible for pollination services to agriculture across large pats of the USA and Europe. These bees have been hit by a condition that is known as Colony Collapse Disorder: in which, as the name implies, colonies collapse to the point where their continued survival is impossible. It is estimated that in some areas the populations of honey bees have declined by 50% in the course of the last decade.

The causes of Colony Collapse Disorder are complex and involve several factors that are not well understood. Bees may be more susceptible if they are exposed to pesticides. However, it seems the major cause is a parasitic mite, called rather dramatically *Varroa destructor*. The mites are quite large relative to the size of their bee hosts, and they infest bees at all life stages, including as they develop in the hexagonal cells of brood combs. What appears to happen is that when the queen bee deposits an egg into the brood cell a small number of female mites will jump into the cell just before it is sealed shut by the worker bees. As the bee larva develops the mites produce offspring and the population in the cell expands ultimately by feeding on the larva itself. This not only kills the larva but generates a large population of mites that can then infect other brood cells. Moreover, the mites not only kill the larvae but they also kill the adult bees because they transmit several deadly bee viruses. It is interesting that while the Western honey bee is spectacularly devastated by infections of Varroa, the Asian honey bee (*Apis cerana*) seems to be able to resist the mite. The reason for the difference between the effects of the mite on the different bee species seems to be that Asian honeybees display a range of behaviours that have been called Varroa hygiene measures. Asian honeybees are able to detect from outside when a brood cell contains mites and when they detect such a cell they uncap it and drag out the larva and kill it. This action prevents the mite population from expanding and causing colony collapse.

Some western honeybees also display the Varroa hygiene behaviour but it is not enough to prevent the mites from causing the colonies to collapse. One approach then to try and solve the Varroa problem is to try and understand what the differences are between honey bees that perform the behaviour and ones that do not. Since we know Asian bees can control the mite by being hygienic, then if western bees could be bred or trained to also exhibit the hygiene behaviour then this might solve the problem. This approach is being pioneered by a group in New Zealand led by Professor Alison Mercer from the University of Otago. In the last couple of years they have published several papers in the journal *Scientific reports*, and this work including some recent unpublished data were presented in a fascinating plenary address at the 22nd International Congress of Zoology held in Okinawa this month (November).

Mercer and colleagues noticed the hygienic bees uncap some cells and kill the larvae, but other cells that are infected remain untouched. Plus of course they do not open and kill uninfected cells. So they must somehow manage to distinguish infected from uninfected cells. How might they do that? There are limited possibilities. The combs are kept in darkness so using visual cues seems unlikely. A good possibility then is that the infected and uninfected cells smell different. To investigate this idea the team collected the odours produced by cells that were infected and those that were not infected. By analysing all the chemicals in the gases given off by each cell type they were able to find a small set of just 5 molecules that were present in the infected cells but absent in the uninfected cells.

It was a good chance that these 5 chemicals were being

used by bees to identify when a cell was infected with mites. To check this out they injected some uninfected cells with the 5 chemicals, and other infected cells with a control set of chemicals that were not enhanced in the infected cells. They predicted that bees would uncap the cells injected with the cocktail of chemicals that they suspected signalled that a cell was infected. As predicted the bees uncapped the chambers where the cocktail of 5 'infested' chemicals were injected, but not the control cells. It appeared then that these chemicals provide a smelly signature of mite infestation.

The key question then was to ask whether the hygiene behaviour was something specific to a subset of the bees, or whether it is a behaviour shown by all bees – but just occasionally. In other words was there something special going on in the bees that were doing the uncapping. Maybe these hygienic bees were able to smell the cocktail of chemicals that signal infestation with mites, but the other bees were not. But how do you tell if a bee can smell something and chooses not to respond, from one that just can't smell it. To test this they watched the bees behaving in the colony and removed a group of bees that had shown the hygiene behaviour. They then put them into a learning task. The task was a kind of bee equivalent of the Pavlov's dog experiment from the 1900's. Pavlov was a

Russian scientist who trained dogs by ringing a bell each time they were fed. The dogs soon associated the ringing of the bell with the food. So eventually when he rang the bell the dogs would salivate because they associated the ringing of the bell with the imminent arrival of food. Now imagine a small refinement of the experiment. Imagine there are 2 bells that ring with different tones. The appearance of food is only paired up with one of the bells. If the dogs can tell the difference between the bells, then when the bells are rung without the appearance of food the dogs would only salivate when they heard the appropriate bell. By making the tones of the bells closer and closer to each other you could actually then work out what the discrimination abilities of the dogs were. That is whether or not they could tell the difference between the bells.

Amazingly bees can be similarly trained using sucrose solution. When a starved bee is given sucrose solution it extends its feeding parts (the proboscis) to drink the solution. So imagine you pair the appearance of sucrose with the appearance of the cocktail of 5 chemicals, but the absence of sucrose with a different cocktail. If the bee is able to distinguish the two sets of chemicals then it will learn only to extend its proboscis (expecting sucrose to come) when the correct chemical set is presented. What

Mercer and colleagues showed was that the bees who had previously been observed uncapping cells were able to distinguish the two sets of chemicals, but bees that did not uncap could not perform the discrimination task. There really did seem to be a difference between the hygienic and unhygienic bees. To explore the cause of this difference the researchers examined the odour receptor populations of the two sets of bees. Odour receptors are molecules in the animals olfaction system that pick up incoming molecules and identify them. They showed that there were significant differences between the two sets of animals. So there is a nice story building up that bees differ in their abilities to smell the infestation signature, and this allows some bees but not others to detect infested cells and uncap them. The missing link is to show that the actual differences in the receptor populations are differences that reflect detection of the critical 5 chemicals.

In theory then if this difference in smell receptor populations between the hygienic and unhygienic bees is genetic it would be possible to breed colonies of bees that have sufficient numbers of individuals displaying the hygiene behaviour to keep the mite populations in check. Great fundamental science addressing an important applied problem. But there is an interesting sting in the tail (excuse the pun). One difference that the researchers observed between the hygienic and unhygienic bees was that the unhygienic ones had greater levels of infection with one of the viruses that is transmitted by the mite! So their inability to smell the infected cells may not be anything to do with their genetics, but actually because the colony is infected with mites in the first place. Ironically then un-infested western bee colonies might consist of mostly hygienic bees that lose their ability just when the colony needs their behaviour the most! In that case solving the problem may prove considerably more complex.

Why do we treat the obese so badly?

December

Each week I get a newsletter which summarises all the publications that appeared during the week on obesity. This week there were 4 separate studies addressing the problem of obesity stigmatization. Individually they are not remarkable, but they add to a growing body of information about a really important problem. Obesity is a strange affliction. People who have obesity suffer terrible prejudice, discrimination and public humiliation about their body weight. The negative bias against people who are obese translates into widespread inequalities across multiple settings, including employment, health care and educational opportunities.

Some of the statistics on discrimination of people because of their body weight are truly shocking. Let's start with the saddest of all. Children are regularly bullied if they are fat. In 2010 a national survey in the USA of overweight sixth grade children (aged around 11-12) revealed that 24% of boys and 30% of girls experienced daily teasing, rejection or bullying directly related to their weight. In high-school (ages 12-17) these numbers increased to 58% of boys and

63% of girls. With approximately 17% of children in the USA now classed as overweight or obese, this translates into millions of children being bullied because of their weight. Because this is sometimes called 'teasing' we downplay its significance. 'Teasing' implies an element of playfulness and humour. Hence, by implication that the person on the receiving end just has no sense of humour if they complain about it, and they are over-reacting. The key however is the intent of the message and the relationship of the two people involved. If one of my friends says 'hey fatty come over here', I know from our overall relationship that they are teasing in a playful way. However, when 'teasing' is not playful it can be extremely hurtful, and a form of bullying. 'Hey fatty, come over here' can then have a much more sinister and hurtful intent when said by a bully to a victim. Indeed this type of action is one of only a wide range of behaviours that come under the mantle of 'bullying', which includes direct physical violence, theft, forcing people to do things against their will, verbal abuse, social rejection and rumour spreading. Imagine the impact of these events happening day in day out, every single day. It is not surprising that occasionally these things get so bad that the children involved cannot take it anymore. There are numerous individual heart rending stories. I will share just one. In 2000 a girl in

Washington DC in the USA was repeatedly bullied and taunted over her weight. Because of this she started to miss school. The school were informed of the reason she was failing to attend by her parents, who asked the school to help, but unfortunately the school did little to try and protect the girl. One school year by February she had already missed over 50 days of school. Eventually rather than help, the school phoned her up to say that unless she attended school immediately she would be put in front of truancy board and face juvenile detention – in effect sent to a juvenile prison. Rather than go back the girl took another route. An hour after the call she hanged herself in her bedroom. It was Valentines day. She was 13. Tragically this is not a one off event, there are lots of other similar stories out there.

Unfortunately, once schooling is over, the prejudice and discrimination against the person with obesity doesn't stop. Most stand-up comedians have parts of their acts that include derogatory statements about obese people. Telling jokes about fat people is completely acceptable in modern societies. If the problem stopped there it would not be too bad, but in fact it extends into much more serious areas. In 2009 a systematic review of the problems faced by the obese adult was published. The review considered several different areas of discrimination but I will present here some of the evidence from just 2 of these areas: the workplace and the healthcare system.

In the workplace there was ample evidence that obese employees received similar negative treatment as obese children. One survey of over 2000 overweight and obese women for example found that 54% reported weight related stigmatization from their co-workers, and 43% reported weight stigma from their actual employers. Examples included being the target of jokes and pejorative comments. But the treatment did not stop at name calling. Many reported that they had been discriminated against by being passed over for promotion, not hired for jobs, or even in the past had been sacked because of their weight. A different study, that also had over 2000 participants, found that overweight respondents were 12x, obese respondents 37x and severely obese respondents 100x more likely than non-obese employees to report employment discrimination, mostly including failure to be hired, or not being promoted. There have been several legal cases where obese individuals have been sacked because of their weight, despite otherwise positive evaluations of their performance, and weight not being an issue with respect to performance of their jobs.

Studies have consistently shown that fat people also earn less. In one study of almost 12,000 subjects it was shown that in men the salary penalty of being obese ranged from 0.7 to 3.4%. In women the problem was even greater with a wages penalty for obese women being 2.3 to 6.1%. A study of over 25,000 people in the USA found that for white females, every 10 pounds of weight (about 5 kg) above normal weight was associated with a 1.5% drop in income. Putting on 30kg of body weight was enough to offset over 3 years of college education! This discrimination is not confined to the USA. In Europe a large study including over 50,000 individuals showed that for every 10% increase in average BMI, income was reduced by 1.9% in males and 3.3% in females. Much of the work in this area involves self-report and correlations, and so it could be that obese people are on average less educated or have other problems that cause both their obesity and their lower earnings. However, there are also many experimental studies. Typical protocols involve giving a panel a CV of a potential candidate, and varying only the photograph of the candidate to show either an obese or lean applicant. In these experiments prospective employer panels are often less keen to employ obese candidates, despite their identical qualifications, personal statements etc. When asked to justify their selections, individuals on the selection panels said that they thought the obese candidates were less conscientious, less agreeable, less emotionally stable, and less extraverted.

If there is one area where you might anticipate that people with obesity might get a sympathetic response it is in the provision of health care. One would think that health professionals might have a more informed view of the obese state. However, far from it. Study after study has shown that physicians have extremely negative views about obese patients classically regarding them as awkward, unattractive, ugly and non-compliant with treatment prescriptions. A study of 600 general practitioners (GPs) in France found that they regarded their obese and overweight patients as lazy and more self indulgent than normal weight patients, and identified lack of will power as the most common problem in treating them. Similar studies in Australia and Israel indicated that GPs regarded lack of motivation and compliance issues to be the main problems trying to treat the obese patient. There is evidence that GPs simply dislike treating obese people. In one study physicians reported that seeing obese patients was a waste of their time, and that heavier patients were more annoying than patients with lower body weights.

This probably explains why medics can often be directly insulting to their obese patients. The UK comedian Phil Hammond, who was formerly a GP in the UK, revealed on TV once that extremely obese patients seeking treatment in the national health service in the UK would have the initials 'D.T.S.' written in big letters on the front of their case notes: which he said was short for 'Danger to Shipping'! In fact derogatory humour directed at obese patients by health care professionals is not uncommon. In 2007 a study involving 54 medical students revealed that severely obese patients were the most common target of derogatory humour by attending physicians and residents.

Students indicated that their denigration of obese patients was due to both the assumption that patients were to blame for their own condition and because patients with obesity caused extra work. Most of the students considered derogatory humour directed at obese patients as entirely appropriate. This attitude does not go unnoticed by the patients. A study which examined the experiences of weight stigma among overweight and obese women found that 53% reported receiving inappropriate comments from doctors about their weight. In addition, doctors were reported as the second most common source of weight stigmatization.

The interesting thing is why? Why do individuals with obesity get treated this way? We would not, for example, ever accept this sort of treatment being handed out to people with cancer. If a stand up comedian was to write a routine about patients with cancer undergoing chemotherapy and losing their hair, my guess is they wouldn't get many laughs. If individuals were passed over for promotion because they had cancer, or they were the butt of derogatory humour by health professionals, there would likely be public outrage. Why is there no outrage when these things are done to obese people?

It is clear people regard cancer and obesity as fundamentally different things. We see cancer as largely a tragedy that is a matter of really bad luck. It could happen to anyone – including ourselves. Most importantly it is a tragedy delivered on people that is not their fault. People with cancer are viewed with sympathy because they are blameless in their health predicament. Obesity is different, because it is widely seen as the result of lifestyle choices that the person has made themselves. These lifestyle choices include being lazy, being greedy and having very little willpower and self control. The consensus of opinion is that obese people are fat because they brought it on themselves. As the comedian Billy Connolly once said 'obese people don't have a metabolism problem – they have a pie eating problem'. People who are not obese generally believe that the reason they are not obese is because they have greater self control that has enabled them to make better lifestyle choices. There is a moral superiority that they feel gives them the permission to judge the obese and make fun of them. This viewpoint extends from comedians to employers and health care professionals.

Study after study confirms that health professionals regard the obese as lazy, stupid, worthless and not motivated to comply with the treatments offered to them. These views simply conform to the wider stereotypes in society that the obese person is lazy, unmotivated, lacking in self-discipline, less competent, less conscientious, less emotionally stable, less agreeable, noncompliant, and sloppy. Yet actually there is no evidence to support these characterisations. For example, a study in the USA of 3176 adults showed that while personality was strongly related to both age and gender it was not associated in any consistent way, and far less significantly, with differences in body weight. Moreover, despite common beliefs among health care professionals that obese patients are unmotivated and non-compliant, actual data suggests obese patients are highly motivated to have their health problems treated. Unfortunately, because of the negative stigma they receive from health professionals they are more likely to delay seeking treatment. These delays can be fatal. Doctor behaviour is literally killing obese patients.

The reason that I raised the issue of cancer, is that in fact obesity and cancer are very similar health problems. In both cases there is a high genetic component and a lifestyle component. With most cancers there is a genetic risk, that is about 70% of the cause. The remainder is due to lifestyle choices. We are only just starting to fill in the complete picture, but smoking and eating processed meat, are well understood lifestyle choices that contribute to the remaining causality of some cancers. These choices have a component that is under voluntary control. We choose to eat processed meat like sausages, and to smoke, despite knowing their health consequences. In obesity the genetic contribution to the overall risk is similarly about 65%. It is important to acknowledge that people that are obese do eat more food than lean people. The key question is why? The answer seems to be little to do with willpower and lack of self control, as the person who is lean normally assumes. In the last 10 years we have made enormous strides forwards in understanding the genetic basis of obesity. In total we now know over 115 genes where variants between individuals are associated with obesity. These genes that we know cause obesity seem to be mostly involved in appetite control in our brains. Obese people have a genetic drive to eat more food and are less easily satiated by food consumption. This is not an issue of will power. In fact obese people very often exert considerably more will power and control with respect to their eating than do lean people. If you are lean and you want to know what this is like then tomorrow morning when you get up skip your breakfast and then miss lunch. By the time you come to eat your evening meal your physiological hunger system will probably be in a state similar to that of an obese person. That is their normal drive to eat. Now try and control it. It isn't easy to overcome these basic physiological urges. Now imagine that you felt like that every time you were going to eat a meal. Would you still have your spectacular willpower?

For sure there is an environmental part as well – which accounts for the remaining 35% of the variation. But this again mostly isn't to do with willpower, or laziness, or lack of self control. The major factors influencing this variation are known to be poverty, socioeconomic class, race and education. Obese people have a physiology that drives them to overconsume, and they make poor food choices most often because they are poor and the cheapest foods are the least healthy. Obviously, some of the food choices obese people make are under voluntary control. It would be incorrect to deny that. However, the importance of voluntary control and willpower is substantially less than most people realise.

A major issue is that although we have made enormous strides in the last 25 years in our understanding of this complex disorder, and despite the fact that obesity is probably the greatest health threat to most western nations, the time that is spent on it in most medical curricula is pitifully small. The result is that most clinicians and general practitioners are no more educated about the causes of obesity than the average prejudiced member of the general public. And the result is they act accordingly. Sooner or later this will need to change and perhaps when it does we will start to make real inroads into the prevailing 'danger to shipping' mentality. Perhaps when everyone sees that medics have started to treat obesity more seriously they will follow suit. It will likely be a long road, but eventually I think we will come to see obesity more and more like cancer, and we will eventually give people who suffer from it the respect they deserve.

Original sources and further reading

January — Do animals keep fit?

Halsey, L.G. (2016) Do animals exercise to keep fit? *Journal of Animal Ecology* **85**: 614-620.

Speakman, J.R. and Krol, E. (2010) Maximal heat dissipation capacity and hyperthermia: neglected key factors in the ecology of endotherms. *Journal of Animal ecology* **79**: 726-46

February — Is democracy the best system of government, or the cradle of mediocrity?

Nagel, M. (2010) A mathematical model of democratic elections. *Current Research: Journal of social sciences* **2**: 255-261

March — Could killing this one type of cell make you live 20% longer?

Hayflick, L. (1965) Limited *in vitro* lifetime of human diploid cell strains. *Experimental cell research* **37**: 614

Campesi, J. and d'adda di Fagagna, F. (2007) Cellular senescence: when bad things happen to good cells. *Nature reviews: Molecular cell Biology* **6:** 729-740.

Baker, D.J. *et al.* (2016) Naturally occurring p16(Ink4a)-positive cells shorten healthy lifespan. *Nature* **530**: 184-187

Van Deursen, J.M. (2014) The role of senescent cells in ageing. *Nature* **509**: 439-446.

April — Why do ant societies tolerate lazy individuals?

Hasegawa, E. *et al.* (2016) Lazy workers are essential for the long term sustainability of insect societies. *Scientific reports*. **6**:20846

May — Shrinking birds

Drent, R.H. and Daan, S (1980) The prudent parent: energetic adjustments in avian breeding. *Ardea* **68**:225-252.

van Gils, J.A. *et al* (2016) Body shrinkage due to arctic warming reduces red knot fitness in tropical wintering range. *Science* **352:** 819-821.

June — Women and children first, or last?

Feldman Hall, O. *et al.* (2016). Moral chivalry: Gender and harm sensitivity predict costly altruism. *Social Psychological & Personality Science,* **7**:542-551

Elinder, M. and Erixson, O. (2012) Gender, social norms and survival in maritime disasters. *Proceedings of the national academy of the USA* **109**: 13220-13224.

July	Why do we not feel hungry when we are ill?

Francesconi, W. *et al.* (2016) The proinflammatory cytokine interleukin 18 regulates feeding by acting on the bed nucleus of the stria terminalis. *Journal of Neuroscience* **36**: 5170-80.

August	Can we make our fat burn itself?

Speakman, J.R. and Heidari-Bakavoli, S. (2016) Type 2 diabetes, but not obesity, prevalence is positively associated with ambient temperature. *Scientific reports* **6**: 30409

September	If you close your eyes how do you know where your legs are?

Chesler, A.T. *et al.* (2016) The role of PIEZO2 in human mechanosensation. *New England Journal of Medicine* **375**: 1355-1364.

October   No more 'Bad air'

Ihizuka, A.S. *et al.* (2016) Protection against malaria at 1 year and immune correlates following PfSPZ vaccination. *Nature medicine* **22:** 614-618.

November   Smelling bee

Mondet F. et al. (2016) Specific cues associated with honey bee social defence against *Varroa destructor* infested brood. *Scientific reports* **6:** 25444

Mondet F. et al. (2015) Antennae hold the key to Varroa sensitive hygiene behaviour in honey bees. *Scientific reports* **5:** 10454

December   Why do we treat the obese so badly?

High, B. (2012) Bullycide in America: Moms speak out about the bullying/suicide connection. Createspace publishing

Marr, N. and Field, T. (2011) Bullycide: death at playtime. 2nd edition. BeWrite Books Ltd

Puhl, R.M. and Heuer, C.A. (2009) The stigma of obesity: a review and update. *Obesity* **17**: 941-964.